W9-BNS-658

More Praise for

SMART CITIES

"An ambitiously wide-ranging, admirably clear-eyed, and ultimately humanistic guidebook to the connected city."

—Tom Vanderbilt, author of *Traffic*

"Compelling. . . . Townsend begins a conversation."

—Melanie E. Moses, *Nature*

"Anthony Townsend's terrific book looks at the historic relationship of urban and industrial development to new technologies."

—*Architecture Today*

"Powerful, readable prose." —Franklyn Cater, NPR

"Townsend's interest in smart cities is more than merely technological: he offers an entertaining history of urban planning's visionaries and villains, the technological breakthroughs and the spectacular failures that brought us to this crossroads."—Tim Smedley, *New Scientist*

"How tomorrow's open spaces evolve cannot be known but armed with this book, the reader will be bang up to date with who's who in the smart city boom, and what's happening where."

—*Engineering and Technology* magazine

"Although the omnipresent surveillance that accompanies this interconnectivity may make some readers nervous, Townsend persuasively demonstrates how ubiquitous information resources can provide more protection, as it did in the Boston marathon bombing case, and facilitate a more comfortable, less stress-inducing city-living experience." —Carl Hays, *Booklist*

"A realistic if dry vision of tomorrow's new breed of cities."
—*Discover*

"Authoritative, information-packed must-read for urban policymakers." —*Kirkus Reviews*, starred review

"Readers of social and urban issues alike will find this a fascinating consideration of smart technology's changes." —*Midwest Book Review*

SMART CITIES

SMART CITIES

Big data,
civic hackers,
and the quest for
a new utopia

Anthony M. Townsend

W. W. NORTON & COMPANY

New York | London

Printed in the United States of America
First published as a Norton paperback 2014

For information about permission to reproduce selections from this book,
write to Permissions, W. W. Norton & Company, Inc.,
500 Fifth Avenue, New York, NY 10110

For information about special discounts for bulk purchases, please contact
W. W. Norton Special Sales at specialsales@wwnorton.com or 800-233-4830

Manufacturing by RR Donnelley
Book design by Chris Welch
Production manager: Louise Mattarelliano

Library of Congress Cataloging-in-Publication Data

Townsend, Anthony M., 1973–
Smart cities : big data, civic hackers, and the quest for a new utopia / Anthony M. Townsend.
— First edition.
pages cm
Includes bibliographical references and index.
ISBN 978-0-393-08287-6 (hardcover)
1. Cities and towns—History. 2. City planning—Technological innovations. 3. Regional planning—Technological innovations. 4. Technological innovations—Economic aspects. 5. Information technology—Economic aspects. I. Title.
HT119.T65 2013
307.76—dc23
2013012755

ISBN 978-0-393-34978-8 pbk.

W. W. Norton & Company, Inc.
500 Fifth Avenue, New York, N.Y. 10110
www.wwnorton.com

W. W. Norton & Company Ltd.
Castle House, 75/76 Wells Street, London W1T 3QT

3 4 5 6 7 8 9 0

For Stella and Carter:

may you thrive in a better world.

What is the city but the people?

—*William Shakespeare,* The Tragedy of Coriolanus

Contents

Preface

Stroll through any neighborhood today and your body sets in motion machines of every kind. Approach a building and the front door slides open. Enter an empty room and a light flicks on. Jump up and down and a thermostat fires up the air conditioner to compensate for the warming air around you. Roam at will and motion-sensing surveillance cameras slowly turn to track you. Day after day, these automatic electromechanical laborers toil at dumb and dirty jobs once done by people. At the fringe of our awareness, they control the world around us. At times they even dare to control us. Yet they are now so familiar, so mundane, that we hardly notice.

But lately these dumb contraptions are getting a lot smarter. Hints of a newly sentient world lurk everywhere. A traffic signal sprouts a stubby antenna and takes its cue from a remote command center. The familiar dials of your electric meter have morphed into electronically rendered digits, its ancient gear works supplanted by a powerful microprocessor. Behind the lens of that surveillance camera lurks a ghost in the machine, an algorithm in the cloud analyzing its field of view for suspicious faces. But what you can see is just the tip of an iceberg. The world is being kitted out with gadgets like these, whose purpose is unclear to the untrained eye. With an unblinking stare, they sniff, scan, probe, and query.

The old city of concrete, glass, and steel now conceals a vast underworld of computers and software. Linked up via the Internet, these devices are being stitched together into a nervous system that supports the daily lives of billions in a world of huge and growing cities. Invisibly, they react to us, rearranging the material world in a flurry of communiqués. They dispatch packages, elevators, and ambulances. Yet, as hectic as this world of automation is becoming, it has a Zen-like quality too. There's a strange new order. Everything from traffic to text messages seems to flow more smoothly, more effortlessly, more in control.

That machines now run the world on our behalf is not just a technological revolution. It is a historic shift in how we build and manage cities. Not since the laying of water mains, sewage pipes, subway tracks, telephone lines, and electrical cables over a century ago have we installed such a vast and versatile new infrastructure for controlling the physical world.

This digital upgrade to our built legacy is giving rise to a new kind of city—a "smart" city. Smart cities are places where information technology is wielded to address problems old and new. In the past, buildings and infrastructure shunted the flow of people and goods in rigid, predetermined ways. But smart cities can adapt on the fly, by pulling readings from vast arrays of sensors, feeding that data into software that can see the big picture, and taking action. They optimize heating and cooling in buildings, balance the flow of electricity through the power grid, and keep transportation networks moving. Sometimes, these interventions on our behalf will go unnoticed by humans, behind the scenes within the wires and walls of the city. But at other times, they'll get right in our face, to help us solve our shared problems by urging each of us to make choices for the greater good of all. An alert might ask us to pull off the expressway to avert a jam, or turn down the air conditioner to avoid a blackout. All the while, they will maintain a vigilant watch over our health and safety, scanning for miscreants and microbes alike.

But the real killer app for smart cities' new technologies is the survival of our species. The coming century of urbanization is humanity's last attempt to have our cake and eat it too, to double down on industrialization, by redesigning the operating system of the last century to cope with the challenges of the coming one. That's why mayors across the globe are teaming up with the giants of the technology industry. These companies—IBM, Cisco, Siemens, among others—have crafted a seductive pitch. The same technology that fueled the expansion of global business over the last quarter-century can compute away local problems, they say. If we only let them reprogram our cities, they can make traffic a thing of the past. Let them replumb our infrastructure and they will efficiently convey water and power to our fingertips. Resource shortages and climate change don't have to mean cutting back. Smart cities can simply use technology to do more with less, and tame and green the chaos of booming cities.

Time will be the judge of these audacious promises. But you don't have to take it sitting down. Because this isn't the industrial revolution, it's the information revolution. You are no longer just a cog in a vast machine. You are part of the mind of the smart city itself. And that gives you power to shape the future.

Look in your pocket. You already own a smart-city construction kit. The democratization of computing power that started with the PC in the 1970s and leaped onto the Internet in the 1990s is now spilling out into the streets. You are an unwitting agent in this historic migration. Stop for a second to behold the miracle of engineering that these hand-held, networked computers represent—the typical CPU in a modern smartphone is ten times more powerful than the Cray-1 supercomputer installed at Los Alamos National Laboratory in 1976. Today, more than 50 percent of American mobile users own a smartphone.[1] Countries all around the world have either already passed, or are fast approaching, the same tipping point.

We are witnessing the birth of a new civic movement, as the smartphone becomes a platform for reinventing cities from the bottom up. Every day, all across the globe, people are solving local problems using this increasingly cheap consumer technology. They are creating new apps that help us find our friends, find our way, get things done, or just have fun. And smartphones are just the start—open government data, open-source hardware, and free networks are powering designs for cities of the future that are far smarter than any industry mainframe. And so, just as corporate engineers fan out to redesign the innards of the world's great cities, they're finding a grassroots transformation already at work. People are building smart cities much as we built the Web—one site, one app, and one click at a time.

SMART CITIES

Introduction

Urbanization and Ubiquity

In 2008, our global civilization reached three historic thresholds. The first came in February when United Nations demographers predicted that within the year, the millennia-long project of settling the planet would move into its final act. "The world population will reach a landmark in 2008," they declared; "for the first time in history the urban population will equal the rural population of the world."[1] We would give up the farm for good, and become a mostly urban species.

For thousands of years, we've migrated to cities to connect. Cities accelerate time by compressing space, and let us do more with less of both. They are where jobs, wealth, and ideas are created. They exert a powerful gravitational pull on the young and the ambitious, and we are drawn to them by the millions, in search of opportunities to work, live, and socialize with each other. While in the end it took slightly longer than the original forecast, by the spring of 2009, most likely in one of China's booming coastal cities or the swelling slums of Africa, a young migrant from the hinterlands stepped off a train or a jitney and tipped the balance between town and country forever.[2]

Cities flourished during the twentieth century, despite humanity's best efforts to destroy them by aerial bombardment and suburban sprawl. In 1900, just 200 million people lived in cities, about one-eighth of the world's population at the time.[3] Today, just over a century later, 3.5 billion call a city home. By 2050, United Nations projections indicate, the urban population will expand to nearly 6.5 billion.[4] By 2100, global population could top 10 billion, and cities could be home to as many as 8 billion people.[5]

This urban expansion is the biggest building boom humanity will ever undertake. Today, India needs to build the equivalent of a new Chicago every year to keep up with demand for urban housing.[6] In 2001, China's announced plans to build twenty new cities each year through 2020, to accommodate an estimated 12 million migrants arriving annually from rural areas.[7] Already largely urbanized, Brazil will instead spend the twenty-first century rebuilding its vast squatter cities, the favelas. In sub-Saharan Africa, where 62 percent of city dwellers live in slums, the urban population is projected to double in the next decade alone.[8] Just in the developing world, it is estimated that one million people are born in or migrate to cities every single week.[9]

The next step was to untether ourselves from the grid. In 2008, for the first time, the number of Internet users who beamed their bandwidth down over the airwaves surpassed those who piped it in over a cable. In the technical jargon of telecommunications industry statisticians, the number of mobile cellular broadband subscribers surpassed the number of fixed DSL, cable, and fiber-optic lines.[10] This shift is being driven by the rapid spread of cheap mobile devices in the developing world, where the mobile web has already won.[11] In India the volume of data sent across wireless networks now surpasses what's conveyed by wire.[12]

Smartphones in hand—over a billion worldwide by 2016, according to Forrester, a market research firm—we are reorganizing our lives and our communities around mass mobile communications.[13] Talking on the go is hardly a new idea—the first mobile phone call was placed in the United States in 1946. But it wasn't until the 1990s that personal mobility came to so dominate and define our lives and demand a telecommunications infrastructure that could keep up. By freeing us to gather where we wish, our mobiles are a catalyst for density; the most robust cellular networks are those that blanket stadiums in bandwidth so spectators can share every score by talking, texting, and photos sent to the social web. But these same networks can be a substrate for sprawl, a metropolitan nervous system conveniently connecting our cars to the cloud. They may be our most critical infrastructure, and seem to be our highest priority. Even as we struggle to find the public will to fund basic maintenance for crumbling roads and bridges, we gladly line up to hand over hard-earned cash to our wireless carriers. Flush with funds, the US wireless industry pumps some $20 billion a year into network construction.[14] While the capital stock invested in the century-old power grid is estimated at $1 trillion in North America alone, nearly $350 billion has been spent in the last twenty-five years on the 285,000 towers that blanket American cities with wireless bandwidth.[15]

The transition away from wires is almost complete. Mobile phones are the most successful consumer electronic device of all time. Some 6 billion are in service around the globe. Three-quarters are in the developing world. In just a few years, it will be unusual for a human being to live without one.

The final transformation of 2008 caught us by surprise. The urban inflection point and the ascendance of wireless were two trends demographers and market watchers had long seen approaching. But just as we verged on linking all of humanity to the global mobile web, we became a minority online. We'll never know what tipped the balance—perhaps a new city bus fired up its GPS tracker for the first time, or some grad students at MIT plugged their coffee pot into Facebook. But at some point the Internet of people gave way to the Internet of Things.[16]

Today, there are at least two additional things connected to the Internet for every human being's personal device. But by 2020 we will be hopelessly outnumbered—some 50 billion networked objects will prowl the reaches of cyberspace, with a few billion humans merely mingling among them.[17] If you think banal chatter dominates the Web today, get ready for the cacophony of billions of sensors tweeting from our pockets, the walls, and city sidewalks, reporting on minutiae of every kind: vehicle locations, room temperatures, seismic tremors, and more. By 2016, the torrent of readings generated by this Internet of Things could exceed 6 petabytes a year on our mobile networks alone (one petabyte equaling one million gigabytes).[18] It will drown out the entire human web—the 10 million photos currently archived on Facebook total a mere 1.5 petabytes.[19] Software in the service of businesses, governments, and even citizens will tap this pool of observations to understand the world, react, and predict. This "big data," as it is increasingly known, will be an immanent force that pervades and sustains our urban world.

This crowded and connected world isn't our future—we are already living in it. Comparing today's China to his first glimpses of the communist state in the 1980s, US ambassador Gary Locke captured the historic nature of this shift. "Now . . . it is skyscrapers, among the tallest in the world," he told PBS talk-show host Charlie Rose on the air in early 2012. "It is phenomenal growth . . . using smartphones everywhere you go. The transformation is just astounding."[20]

But the transformation is just getting started. This book explores the intersection between urbanization and the ubiquitous digital technology that will shape our world and how we will live in it. How we guide the integration of these historic forces will, to a great extent, determine the kind of world our children's children will inhabit when they reach the other end of this century. But before we look ahead, it makes sense to look back. For this is but the last act in a drama that has played out since the beginning of civilization.

Symbiosis

The symbiotic relationship between cities and information technology began in the ancient world. Nearly six thousand years ago, the first markets, temples, and palaces arose amid the irrigated fields of the

Middle East and served as physical hubs for social networks devoted to commerce, worship, and government. As wealth and culture flourished, writing was invented to keep tabs on all of the transactions, rituals, and rulings. It was the world's first information technology.

In more recent eras, each time human settlements have grown larger, advances in information technology have kept pace to manage their ever-expanding complexity. During the nineteenth century, industrialization kicked this evolutionary process into high gear. New York, Chicago, London, and other great industrial cities boomed on a steady diet of steam power and electricity. But this urban expansion wasn't driven only by new machines that amplified our physical might, but also by inventions that multiplied our ability to process information and communicate quickly over great distances. As Henry Estabrook, the Republican orator (and attorney for Western Union) bombastically declared in a speech honoring Charles Minot, who pioneered the use of the telegraph in railroad operations in 1851, "The railroad and the telegraph are the Siamese twins of Commerce, born at the same period of time, developed side by side, united by necessity."[21]

The telegraph revolutionized the management of big industrial enterprises. But it also transformed the administration of city government. Police departments were among the earliest adopters, using the tool to coordinate security over growing jurisdictions.[22] Innovations flowed from government to industry as well—the electromechanical tabulators invented to tally the massive 1890 census were soon put to use by corporations to track the vital signs of continent-spanning enterprises. By enabling business to flourish and municipalities to govern more effectively, these technologies removed critical obstacles to the growth of cities. By 1910, historian Herbert Casson could declare matter-of-factly what was clear to all about yet another technology. "No invention has been more timely than the telephone," he wrote. "It arrived at the exact period when it was needed for the organization of great cities and the unification of nations."[23]

For anyone who has telecommuted to work or watched a live broadcast from the other side of the planet, it seems counterintuitive that the growth of cities and the spread of information technology are so strongly linked. Many have argued the opposite—that new technologies undermine the need for cities and all of the productive yet expensive and sometimes unpleasant proximity they provide. In 1964 science-fiction legend Arthur C. Clarke articulated a vision of the future where, thanks to satellite communications, "It will be possible . . . perhaps only fifty years from now, for a man to conduct his business from Tahiti or Bali, just as well as he could from London."[24] More recently, as the Internet began its meteoric rise in the mid-1990s, tech pundit George Gilder wrote off cities as "leftover baggage from the industrial era."[25] But instead of disintegrating, London grew bigger, richer, more vital and connected than ever. Instead of undermining the city, new telecommunications technologies played a crucial role in London's success—it is the hub of a global tangle of fiber-optic networks that plug its financiers and media tycoons directly into the lives of billions of people all over the world.

We experience the symbiosis of place and cyberspace everyday. It's almost impossible to imagine city life without our connected gadgets. In my own pocket, I carry an iPhone. It is my megacity survival kit, a digital Swiss Army knife that helps me search, navigate, communicate, and coordinate with everyone and everything around me. I have apps for finding restaurants, taxis, and my friends. A networked calendar keeps me in sync with my colleagues and my family. If I'm running late, there are three different ways to send a message and buy some time. But I'm not alone. We've all become digital telepaths, hooked on the rush we get as these devices untether us from the tyranny of clocks, fixed schedules, and prearranged meeting points. The addiction started, as all do, slowly at first. But now it governs the metabolism of our urban lives. With our days and nights increasingly stretched across the vastness of megacities, we've turned to these smart little gadgets to keep it all synchronized. It's no accident that the most common text message, sent billions of times a year all over the world, is "where r u?"[26]

The digital revolution didn't kill cities. In fact, cities everywhere are flourishing because new technologies make them even more valuable and effective as face-to-face gathering places.

Struggle

Beginning in the 1930s, men like Robert Moses began rebuilding cities around a new technology, the automobile. Moses was an autocrat and technocrat, a master planner and "power broker" (the title of Robert Caro's epic biography). His disdain for the accumulated architectural canvas he inherited was no secret. "You can draw any kind of picture you like on a clean slate and indulge your every whim in the wilderness of laying out a New Delhi, Canberra or Brasilia," he said of the new capital cities of that era, "but when you operate in an overbuilt metropolis you have to hack your way with a meat ax."[27] For three decades, in various public posts in New York and elsewhere as a consultant, Moses brought to life the dazzling vision of a middle-class, motorized America first unveiled by General Motors at the 1939 World's Fair in New York City. To make way for the future, he bulldozed the homes of over a quarter-million unfortunate New Yorkers.[28]

Today, a new group of companies have taken GM's spot in the driver's seat and are beginning to steer us toward a new utopia, delivered not by road networks but by digital networks. Instead of paving expressways through vibrant neighborhoods, these companies hope to engineer a soft transformation of cities through computing and telecommunications. "Drivers now see traffic jams before they happen," boasts an IBM advertisement posted in airports all over the world. "In Singapore, smarter traffic systems can predict congestion with 90% accuracy." With upgrades like these, unlike Moses, we may never need to pave another mile of roadway.

For the giants of the technology industry, smart cities are fixes for the dumb designs of the last century to prepare them for the challenges of the next, a new industrial revolution to deal with the unintended consequences of the first one. Congestion, global warming, declining health—all can simply be computed away behind the scenes. Sensors, software, digital networks, and remote controls will automate the things we now operate manually. Where there is now waste, there will be efficiency. Where there is volatility and risk, there will be predictions and early warnings. Where there is crime and insecurity, there will be watchful eyes. Where you now stand in line, you will instead access government services online. The information technology revolution of the nineteenth century made it possible to govern industrial cities as their population swelled into the millions. This revolution hopes to wrest control over cities of previously unthinkable size—ten, twenty, fifty, or even one hundred million people.

With a potential market of more than $100 billion through the end of this decade, many of the world's largest companies are jockeying for position around smart cities.[29] There are the engineering conglomerates that grew to greatness building the systems that control our world: IBM, which sprang from the company that built the tabulators for the 1890 census; Siemens, which got its start by wiring up German cities with telegraph cables; and General Electric, which lit up America's cities with artificial light. But there are newcomers, too, like Cisco Systems, the master plumber of the Internet. For each, success in selling us on smart cities will pave the way for decades of growth. Peering out from the cover of *Forbes* in 2011, CEO Peter Löscher of Siemens summed up the hopes of corporate leaders everywhere as he gushed at the prospect of supplying infrastructure for the cities of the developing world, "This is a huge, huge opportunity."[30]

By the 1970s, the construction of urban expressways in the United States had ground to a halt, stopped by a grassroots rebellion that held very different views of the role of cars, how city planning should be conducted, and even the very nature of the city itself. The first signs of a similar backlash to corporate visions of smart cities are now coming to light, as a radically different vision of how we might design and build them bubbles up from the street. Unlike the mainframes of IBM's heyday, computing is no longer solely in the hands of big companies and governments. The raw material and the means of producing the smart city—smartphones, social software, open-source hardware, and cheap bandwidth—are widely democratized and inexpensive. Combining and recombining them in endless variations is cheap, easy, and fun.

All over the world, a motley assortment of activists, entrepreneurs, and civic hackers are tinkering their ways toward a different kind of utopia. They eschew efficiency, instead seeking to amplify and accelerate the natural sociability of city life. Instead of stockpiling big data, they build mechanisms to share it with others. Instead of optimizing government operations behind the scenes, they create digital interfaces for people to see, touch, and feel the city in completely new ways. Instead of proprietary monopolies, they build collaborative networks. These bottom-up efforts thrive on their small scale, but hold the potential to spread virally on the Web. Everywhere that industry attempts to impose its vision of clean, computed, centrally managed order, they propose messy, decentralized, and democratic alternatives.

It's only a matter of time before they come to blows.

Experimentation

At the middle of this emerging battlefield sits City Hall. Encamped on one flank are industry sales teams, proffering lump sums up front in return for exclusive contracts to manage the infrastructure of cash-strapped local governments. On the other flank, civic hackers demand access to public data and infrastructure. But even as they face the worst fiscal situation in a generation—in the United States, in Europe, even in China—cities are rapidly emerging as the most innovative and agile layer of government. Citizens routinely transcend the tyranny of geography by going online, but local governments are still the most plugged in to their daily concerns. Yet citizen expectations of innovation in public services continue to grow, while budgets shrink. Something has to give.

For a new cadre of civic leaders, smart technology isn't just a way to do more with less. It's a historic opportunity to rethink and reinvent government on a more open, transparent, democratic, and responsive model. They are deploying social media to create more responsive channels of communication with citizens, publishing vast troves of government data on the Web, and sharing real-time feeds on the location of everything from subways to snowplows. There's also a huge economic opportunity. By unlocking public databases and building broadband infrastructure, many cities hope to spawn homegrown inventions that others will want to buy, and attract highly mobile entrepreneurs and creative talent. Looking smart, perhaps even more than actually being smart, is crucial to competing in today's global economy.

Zoom out from the local to the global scale and, like a satellite photo of the earth at night, a twinkling planet of civic laboratories comes into view. According to Living Labs Global, a Barcelona-based think tank that tracks the international trade in smart-city innovations, there are over 557,000 local governments worldwide.[31] As they begin to experiment with smart technology, each faces a unique set of challenges and opportunities with a different pool of resources. Much as there are mobile apps for every purpose we can imagine, smart cities are being crafted in every imaginable configuration. Local is the perfect scale for smart-technology innovation for the same reasons it's been good for policy innovation—it's much easier to engage citizens and identify problems, and the impact of new solutions can be seen immediately. Each of these civic laboratories is an opportunity to invent.

But each local invention is also an opportunity to share with other communities. For the last few decades, as the pace of globalization accelerated, multinational corporations were the primary means by which technological innovation spread from place to place. Industry would love to play the role of Johnny Appleseed again with smart-city technology. But cities have become highly adept at sharing and copying new innovations on their own, as evidenced in an accelerating diffusion of good ideas. Bus rapid transit, a scheme for improving the capacity of bus lines with dedicated lanes and other clever tweaks, has taken forty years to spread from its birthplace in Curitiba, Brazil, in 1974 to over 120 cities all over the world.[32] Public bike sharing, which surged onto the global stage with the launch of Paris's Vélib system in 2007, has reached a similar footprint in just a few years. Today, there is a bustling trade not just in case studies and best practices of smart-city innovations but actual working technology: code, computer models, data, and hardware designs. These digital solutions can spread quite literally overnight.

The spectacular array of local innovations being cooked up in the world's civic laboratories will challenge our assumptions about both technology and cities, and how they should shape each other. Technologists often want to cut to the chase, find the killer app, and corner the market—this dynamic is already at work in corporate plans for cookie-cutter smart cities. But if we want to get the design of smart cities right, we need to take into account local quirks and involve citizens in their creation. Over time, we'll surely extract the essence of what's reusable and share it widely. But building smart cities is going to take time. It will by necessity be a long, messy, incremental process.

Crash

Every city contains the DNA of its own destruction—some existing fissure that, under pressure, can erupt into conflict or cascade into collapse.

Smart technologies are already fueling conflict between factions in divided cities. The extent of the role played by social media in the

2011 urban uprisings of the Arab Spring has been hotly debated. But Facebook, Twitter, and YouTube were a mere sideshow to the torrent of text messages that turned angry crowds into smart mobs, as they have done numerous times since 2001, when they summoned some 700,000 Filipinos to protests against corrupt President Joseph Estrada. These wireless channels, which provide what is for all intents and purposes a rudimentary form of telepathic communication, were so important that at the height of the Egyptian uprising authorities lobotomized Cairo by ordering a shutdown of the nation's cellular networks. While this act didn't stop the revolution (and probably hastened the flow of remaining bystanders out into the streets), blacking out cities' wireless networks is becoming a disturbingly appealing option for security officials in the West as well—in August 2011 transit police jammed cellular signals during antipolice protests in San Francisco. The same week officials in the United Kingdom discussed blocking the BlackBerry Messenger mobile messaging service and other social media being used to coordinate widespread urban rioting.[33]

Smart cities may also amplify a more commonplace kind of violence—that inflicted by poverty—by worsening gaps between haves and have-nots. This may happen by design, when sensors and surveillance are used to harden borders and wall off the poor from private gated communities. Or it may simply be an unintended consequence of poorly thought-through interventions.

In 2001, the government of India's Karnataka state set out to reform the way it tracked land ownership, ostensibly to root out village-level corruption. Bhoomi, as the new digital recording system was called, was funded by the World Bank as a model for e-government reforms throughout the developing world. But it had the opposite impact. The village-level officials who had administered the old system had always taken bribes, but in return, they interpreted documents for the illiterate and provided advice on how to navigate complex legal procedures. Bhoomi certainly curbed village level corruption—the number of persons reporting paying bribes fell from 66 percent to 3 percent. But centralizing records merely centralized corruption. Wealthy speculators with deep pockets simply targeted officials at higher levels, allowing them to rapidly appropriate land in the expansion path of the region's fast-growing capital, Bangalore.[34] As one development scholar has noted, "While in theory, the initiative was intended to democratize access to information, in practice the result was to empower the empowered."[35] As similar digitization efforts transform government everywhere, the stakes for the poor are enormous. In this new computational arms race, poor communities will be at the mercy of those who can measure and control them from a distance.

Even if there is peace and equality, the smart city may come crashing down under its own weight because it is already buggy, brittle, and bugged, and will only become more so. Smart cities are almost guaranteed to be chock full of bugs, from smart toilets and faucets that won't operate to public screens sporting Microsoft's ominous Blue Screen of Death. But even when their code is clean, the innards of smart cities will be so complex that so-called normal accidents will be inevitable. The only questions will be when smart cities fail, and how much damage they cause when they crash. Layered atop the fragile power grid, already prone to overload during crises and open to sabotage, the communications networks that patch the smart city together are as brittle an infrastructure as we've ever had.

Before it ever comes close to collapse, we might tear down the walls of the smart city ourselves, for they will be the ultimate setup for surveillance. Will smart cities become the digital analogue of the Panopticon, Jeremy Bentham's 1791 prison design, where the presence of an unseen watcher kept order more effectively than the strongest bars?[36] In the 1990s, the Surveillance Camera Players staged sidewalk performances at camera locations in New York City to protest the rapid spread of video monitoring in public spaces. As we install countless new devices that record, recognize, influence, and control our movements and behaviors, this whimsical dissent will seem quaint in retrospection. For as the true value of these technologies for governments and corporations to spy on citizens and consumers alike becomes apparent, the seeds of distrust will bloom. In 2012, concerned about the risks of face-recognition technology, US Senator Al Franken said, "You can change your password, and you can get a new credit card, but you can't change your fingerprint, you can't change your face—unless, I guess, you go through a great deal of trouble."[37] But devious countermeasures are already spreading. In the place of protest, more pragmatic responses are popping up, like Adam Harvey's CV Dazzle. A face-painting scheme based on World War I antisubmarine camouflage, CV Dazzle is designed to confuse face-recognition algorithms.[38]

A New Civics

If the history of city building in the last century tells us anything, it is that the unintended consequences of new technologies often dwarf their intended design. Motorization promised to save city dwellers from the piles of horse manure that clogged nineteenth-century streets and deliver us from a shroud of factory smoke back to nature. Instead, it scarred the countryside with sprawl and rendered us sedentary and obese. If we don't think critically now about the technology we put in place for the next century of cities, we can only look forward to all the unpleasant surprises they hold in store for us.

But that's only if we continue doing business as usual. We can stack the deck and improve the odds, but we need to completely rethink our approach to the opportunities and challenges of building smart cities. We need to question the confidence of tech-industry giants, and organize the local innovation that's blossoming at the grassroots into a truly global movement. We need to push our civic leaders to think more about long-term survival and less about short-term gain, more about cooperation than competition. Most importantly, we need to take the wheel back from the engineers, and let people and communities decide where we should steer.

People often ask me, "What is a smart city?" It's a hard question to answer. "Smart" is a problematic word that has come to mean a million things. Soon, it may take its place alongside the handful of international cognates—vaguely evocative terms like "sustainability" and "globalization"—that no one bothers to translate because there's no consensus about what they actually mean. When people talk about smart cities, they often cast a wide net that pulls in every new public-service innovation from bike sharing to pop-up parks. The broad view is important, since cities must be viewed holistically. Simply installing some new technology, no matter how elegant or powerful, cannot solve a city's problems in isolation. But there really is something going on here—information technology is clearly going to be a big part of the solution. It deserves treatment on its own. In this book, I take a more focused view and define smart cities as places where information technology is combined with infrastructure, architecture, everyday objects, and even our bodies to address social, economic, and environmental problems.

I think the more important and interesting question is, "what do you want a smart city to be?" We need to focus on how we shape the technology we employ in future cities. There are many different visions of what the opportunity is. Ask an IBM engineer and he will tell you about the potential for efficiency and optimization. Ask an app developer and she will paint a vision of novel social interactions and experiences in public places. Ask a mayor and it's all about participation and democracy. In truth, smart cities should strive for all of these things.

There are trade-offs between these competing goals for smart cities. The urgent challenge is weaving together solutions that integrate these aims and mitigate conflicts. Smart cities need to be efficient but also preserve opportunities for spontaneity, serendipity, and sociability. If we program all of the randomness out, we'll have turned them from rich, living organisms into dull mechanical automatons. They need to be secure, but not at the risk of becoming surveillance chambers. They need to be open and participatory, but provide enough support structure for those who lack the resources to self-organize. More than anything else, they need to be inclusive. In her most influential book, *The Death and Life of Great American Cities*, the acclaimed urbanist Jane Jacobs argued that "cities have the capability of providing something for everybody, only because, and only when, they are created by everybody."[39] Yet over fifty years later, as we set out to create the smart cities of the twenty-first century, we seem to have again forgotten this hard-learned truth.

But there is hope that a new civic order will arise in smart cities, and pull every last one of us into the effort to make them better places. Cities used to be full of strangers and chance encounters. Today we can mine the social graph in an instant by simply taking a photo. Algorithms churn in the cloud, telling the little things in our pocket where we should eat and whom we should date. It's a jarring transformation. But even as old norms fade into the past, we're learning new ways to thrive on mass connectedness. A sharing economy has mushroomed overnight, as people swap everything from spare bedrooms to cars, in a synergistic exploitation of new technology and more earth-friendly consumption. Online social networks are leaking back into the thriving urban habitats where they were born in countless promising ways.

These developments are our first baby steps in fashioning a new civics for smart cities. The last chapter of this book lays out the tenets I think can guide us in navigating the decisions we'll make in the coming decade as we deploy these technologies in our communities.

Your Guide

For the last fifteen years, I've watched the struggle over how to build smart cities evolve from the trenches. I've studied and critiqued these efforts, designed parts of them myself, and cheered others along. I've written forecasts for big companies as they sized up the market, worked with start-ups and civic hackers toiling away at the grass roots, and advised politicians and policy wonks trying to push reluctant governments into a new era. I understand and share much of their agendas.

But I've also seen my share of gaps, shortfalls, and misguided assumptions in the visions and initiatives that have been carried forth under the banner of smart cities. And so I'm going to play the roles of myth buster, whistle-blower, and skeptic in one. New technologies inspire us to dream up new ways of living. The promise of technological fixes to complex social, economic, and environmental problems is seductive. Many of the people you will encounter in this book have placed their bet on a better future delivered through technology. Not me. I get nervous when I hear people talk about how technology is going to change the world. I have been around technology enough to know its vast potential, but also its severe limitations. When coarsely applied to complex problems, technology often fails.

What's much more interesting is how we are going to change our technology to create the kinds of places we want to live in. I believe that's going to happen at the grass roots, and I hope my vision of the tremendous resilience and potential for innovation in every city will carry you through the darker moments of this book. I think there is an important role for industry, but my objective here is to put an end to the domination of corporate visions in these early conversations about the future of cities.

Above all, I'm an advocate for cities and the people that live in them. Technology pundits can preach from behind a screen, but cities can't be understood only by looking inside City Hall or a boardroom. You have to connect the schemes of the rich and powerful with the life of the street. That means taking a broad historical and global view of the landscape. To understand the choices we have ahead of us and the unintended consequences, and articulate a set of principles that can better guide our plans and designs moving forward, we need to reexamine how cities and information and communications technologies have shaped each other in the past.

We're also going to skip around. A lot. There isn't any single place we can go to see a smart city in its entirety—they are emerging in fits and starts all across the world. And some of the things we'll see may not be here tomorrow. The smart city is a work in progress. Each day, we lay new wires and mount new antennas, load new software, and collect new data. By the time you read this, many of the technologies described in this book will have evolved. A few will be obsolete. New inventions will have taken their place.

Still, the struggle will remain. The technology industry is asking us to rebuild the world around its vision of efficient, safe, convenient living. It is spending hundreds of millions of dollars to convince us to pay for it. But we've seen this movie before. As essayist Walter Lippmann wrote of the 1939 World's Fair, "General Motors has spent a small fortune to convince the American public that if it wishes to enjoy the full benefit of private enterprise in motor manufacturing, it will have to rebuild its cities and its highways by public enterprise."[40] Today the computer guys are singing the same song.

I believe there is a better way to build smart cities than to simply call in the engineers. We need to lift up the civic leaders who would show us a different way. We need to empower ourselves to build future cities organically, from the bottom up, and do it in time to save ourselves from climate change. This book shows you it can be done, one street corner at a time. If that seems an insurmountable goal, don't forget that at the end of the day the smartest city in the world is the one you live in. If that's not worth fighting for, I don't know what is.

1

The $100 Billion Jackpot

Throughout history, the construction of great gathering spaces has always pushed the limits of technology. The Crystal Palace, a vast, soaring structure of iron and glass built in London's Hyde Park was no exception. The brainchild of Joseph Paxton, a master gardener and architect of greenhouses, the Crystal Palace was a stage for one of the most celebrated international expos of all time, the Great Exhibition of 1851. It was the architectural expression of Victorian England's fast-growing industrial might.

But with industrial-scale architecture came industrial-scale management challenges. As new materials and advances in structural engineering permitted the construction of ever-larger buildings in the nineteenth century, it became more and more difficult to manage the growing flows of people, air, water, and waste that coursed through them each day. With all its glass, the Crystal Palace was, by Paxton's design, a massive greenhouse. Without proper ventilation, the building would have simply cooked the 90,000 visitors its vast expanses could hold.

With the invention of modern air-conditioning still a half-century in the future, Paxton desperately needed a way to boost the building's own natural ventilation. His solution was a system of louvered vents that ran along the building's eaves, which could be opened to release rising hot air and draw in cooler air through the many ground-level entrances. Mechanical rods and levers were fastened into place linking the controls for multiple vents in 300-foot clusters, greatly reducing the labor involved in opening and closing them. Manned by a small team of attendants from the Royal Sappers and Miners, the British military's engineering corps, the vents were adjusted every two hours based on readings from fourteen thermostats placed throughout the structure.[1] While far from automatic, the Crystal Palace's ventilation system showed how mechanical controls and sensors could work together to dynamically reconfigure an entire, massive building in response to changes in the environment. Paxton's contraption was a harbinger of the automation revolution that will transform the buildings and cities we live in over the coming decades.

More than a century later, at the dawn of the computer age, a design for a very different kind of gathering space spurred another bold leap into building automation. Howard Gilman was the heir to a paper-making fortune but his true avocation was philanthropist and patron of the arts. Gilman lavished his family fortune on a variety of causes, supporting trailblazers in dance, photography, and wildlife preservation. In 1976, he began making plans to establish a creative retreat for his network of do-gooders to gather and contemplate a better world.[2] To bring his vision to life, Gilman engaged the English architect Cedric Price.

Price taught at the school of London's Architectural Association, which in the 1960s had spawned the avant-garde Archigram group. In a series of pamphlets, Archigram's members published a variety of hypothetical designs that took new technologies and pushed them to the edge of plausibility. Ron Herron's "Walking City" (1964), the most famous, illustrated a plan for football-shaped buildings propelled by a set of eight insect-like robotic legs.[3] Archigram's fanciful designs were but the latest expression of a long line of architects who were obsessed with movement and the potential of machines to merge with buildings and make them come to life. As American architectural critic Michael Sorkin notes, "The group was squarely a part of a historic British movement visible in a line of engineered structures running through the Crystal Palace, the Dreadnought, the Firth Bridge, the Sopwith Camel, and the E-Type Jag."[4]

For the retreat, to be built at White Oak Plantation, the bucolic family estate on Florida's St. Mary's River, Gilman's design brief was concise but challenging, calling for "A building which will not contradict, but enhance, the feeling of being in the middle of nowhere; has to be accessible to the public as well as to private guests; has to create a feeling of seclusion conducive to creative impulses, yet . . . accommodate audiences; has to respect the wildness of the environment while accommodating a grand piano; has to respect the continuity of the history of the place while being innovative."[5]

Price's response to this set of contradictory demands was "Generator." Less of a building, Generator was more a set of building blocks, 150 stackable 12-foot cubes, "all of which could be moved by mobile crane as desired by users to support whatever activities they had in mind, whether public or private, serious or banal," according to architectural historian Molly Steenson.[6]

But Price worried that people might not take up the challenge of rearranging the building often enough. In the spirit of Archigram's robotic fantasies, Price called on the husband-and-wife team of John and Julia Frazer, architects with deep computer programming expertise, to write software that would do so automatically. The "perpetual architect" program the Frazers created was designed to eliminate boredom. It would sense the layout of the modules and reassemble them overnight into a new pattern to provoke, delight, and otherwise stimulate the retreat-goers. "In the event of the site not being re-organized or changed for some time the computer starts generating unsolicited plans and improvements. . . . In a sense the building can be described as being literally 'intelligent,' " they told Price in a letter. It "should have a mind of its own."[7]

Generator was never built, as concerns about the cost of maintaining the building came to light and Gilman struggled with his younger brother Chris over control of the family fortune.[8] Yet it was an important early vision of how a building—and by extension entire cities—might be transformed by their coming integration with computers. By combining digital sensing, networking, intelligence, and robotics, Price and the Frazers had invented what architect Royston Landau described as "a computerized leisure facility, which not only could be formed and reformed but, through its interaction with users, could learn, remember and develop an intelligent awareness of their needs."[9]

The Automatic City

Economic shocks have an uncanny ability to distill impractical but promising new technologies into commercial successes. Just as Generator was prodding architects to think about computers as architectural materials, the oil embargoes of the 1970s spurred a more prosaic, yet more widespread interest in building automation. "At the time, buildings tended to be over-designed and over-ventilated, and energy efficiency was rarely an issue," notes one industry retrospective.[10] It was clear that a new way of running buildings was needed and automation was the key. Throughout the 1970s and 1980s, energy management systems began appearing in new constructions—simple controls that could adjust heating and cooling controls on a preprogrammed schedule. But as energy costs collapsed in the 1990s, interest in building automation waned, almost as quickly as America's interest in compact, fuel-efficient cars.

Today, high energy costs are back, but the urgency of reducing greenhouse gas emissions is the driving force behind a new surge of investment in building automation. Price's and Frazer's vision of intelligent structures that would adapt to uplift the soul has devolved into something more mundane. The blueprints for smart buildings today co-opt automation merely to sustain the human body on a low-carbon diet. High architectural art has become a tool for cost-cutting and environmental compliance.

This new commercial reality is on display at yet another great gathering space, the Songdo Convensia Convention Center, the hub of a vast new city in South Korea. Rising atop 1,500 acres of landfill reclaimed from the shallows of the Yellow Sea, Songdo International Business District seeks to scale building automation up to an entire city, and cut greenhouse gas emissions by two-thirds.[11]

Convensia's own soaring metal trusses evoke those of the Crystal Palace a century and a half earlier. Overhead they bear the weight of three long, peaked roof sections that enclose one of the largest column-free spans in Asia, according to the building's official website. But behind the scenes, Convensia's true homage to Paxton lies in the control systems that govern every aspect of building function. Here, everything is connected, everything is automated.

Upon entering the building, conventioneers pick up their ID badges, embedded with a "u-chip" (for "ubiquitous" computing), a radio-frequency identification (RFID) tag that functions as a wireless bar code. To enter the exhibition hall, one swipes the card across a reader mounted atop each turnstile, much like entering a subway station. It's a familiar move for Korean city dwellers. For over a decade, they have used local tech giant LG's rechargeable T-money cards not just to board buses and subways, but to pay for taxis and convenience-store purchases as well. From the earliest planning stages, the nation's economic planners intended Songdo to be a test bed for RFID and a center for research and development in this crucial ubiquitous computing technology. In 2005 the government announced a $300 million, 20-acre RFID-focused industrial park in Songdo.[12]

Inside Convensia, your interactions with computers seem far from ubiquitous, broken up into a fragmented series of gestures and glances—swiping your RFID card to enter a room or pressing a button to request that an elevator be dispatched to your location. As they move through the complex, visitors locate meeting rooms by reading digital displays mounted beside entryways, which draw down the latest events schedule from a central master calendar. Other smart technologies inhabit Convensia's unseen innards—controls for climate systems, lighting, safety and security systems are there, yet invisible to the average person.

Step outside, however, and the street springs to life as a less patient, more proactive set of automated technologies takes over. Songdo is the world's largest experiment in urban automation, with millions of sensors deployed in its roads, electrical grids, water and waste systems to precisely track, respond to, and even predict the flow of people and material. According to CEO John Chambers of Cisco Systems, which committed $47 million in 2009 to build out the city's digital nervous system, it is a place that will "run on information."[13] Plans call for cameras that detect the presence of pedestrians at night in order to save energy safely by automatically extinguishing street lighting on empty blocks. Passing automobiles with RFID-equipped license plates will be scanned, just the way conventioneers are at Convensia's main gate, to create a real-time map of vehicle movements and, over time, the ability to predict future traffic patterns based on the trove of past measurements.[14] A smart electricity grid will communicate with home appliances, perhaps anticipating the evening drawdown of juice as tens of thousands of programmable rice cookers count down to dinnertime.

Just above the northern horizon, a line of wide-body jets stretches out over the water, on final approach into the massive Incheon International Airport, which opened in March 2001. The airport is to Songdo what New York's harbor or Chicago's railyards once were. As John Kasarda and Greg Lindsay explain in their 2011 book *Aerotropolis*, Songdo was originally conceived as "a weapon for fighting trade wars." The plan was to entice multinationals to set up Asian operations at Songdo, where they would be able to reach any of East Asia's boomtowns quickly by air. It was to be a special economic zone, with lower taxes and less regulation, inspired by those created in Shenzhen and Shanghai in the 1980s by premier Deng Xiaoping, which kick-started China's economic rise.[15]

But in an odd twist of fate, Songdo now aspires to be a model *for* China instead. The site itself is deeply symbolic. Viewed from the sky, its street grid forms an arrow aimed straight at the heart of coastal China. It is a kind of neoliberal feng shui diagram, drawing energy from the rapidly urbanizing nation just over the western horizon. Massive in its own right, Songdo is merely a test bed for the technology and business models that will underpin the construction of pop-up megacities across Asia. It is the birth of what Michael Joroff of MIT describes as a "new city-building industry," novel partnerships between real estate developers, institutional investors, national governments, and the information technology industry. This ambition to become the archetype for Asia's hundreds of new towns is why scale matters so much for Songdo. Begun in 2004 and scheduled for completion in 2015, it is the largest private real estate project in history at some $35 billion. For Lindsay, it is simply "a showroom model for what is expected to be the first of many assembly-line cities."[16]

South Korea is fertile ground for rethinking the future. It's an anxious place inhabited by driven people, where the phrase *pali pali* is a ubiquitous incantation. Hearing it spoken so often, the foreign ear easily assumes that it is local parlance for "yes" or "please." But it really means "hurry, hurry." It's the verbal expression of the Koreans' approach to most everything, especially city building. No country has industrialized and urbanized as fast and as thoroughly as Korea did during the second half of the twentieth century. In 1953 the country lay in ruins, split in two by a civil war that claimed millions of lives. The citizens of Seoul began rebuilding from near-total destruction. Between 1950 and 1975, the city's population doubled approximately every nine years, growing from just over 1 million people in 1950 to almost 7 million people in 1975. But by the 1990s, according to a report by the Seoul Development Institute, the city's urban-planning think tank, "one could say that Seoul was no longer an independent city but was rather the central city of a rapidly expanding metropolitan region of 20 million."[17] To call Songdo a new "city" is ill conceived—it is merely Seoul's newest and farthest-flung satellite town.

As a test bed for digital technology, Seoul in the early twenty-first century is hard to beat, with over a decade of widespread experience with broadband Internet. After a bailout from the IMF during a financial crisis in 1997, South Korea embraced the Internet as an engine of economic recovery and social transformation. The national government modernized telecommunications laws, invested in a national broadband network, and launched a volley of new policies to push the use of broadband in education, health care, and delivery of government services. From just 700,000 mostly dial-up Internet subscribers in 1997, by 2002 Seoul was home to some 4.5 million broadband households. That year, as plans for Songdo were only just taking shape, one in every twelve broadband Internet users in the industrialized world was living in Seoul, and one in six was Korean. There were more broadband homes in the single city of Seoul than in the entire nations of Canada, Germany, or the United Kingdom. Over twenty thousand Internet cafes, or "PC bangs" (literally, "PC rooms"), had created a broadband culture unlike anything else on earth.[18] The city was unique in the world, a glimpse into a high-speed connected future. Building Songdo was a natural next step. Much as Frank Lloyd Wright's utopian 1932 plan for Broadacre City reimagined a thoroughly suburbanized America around the capabilities of the automobile, Songdo would reimagine the Korean metropolis around the potential of ubiquitous computing. It was, in fact, the first of a series of "u-cities" conceived by the national government to make Korea a world leader in smart-city technology and construction.

Korea is a prosperous nation, but Songdo was also an expression of anxiety about the rise of modern China, and the threats it would pose for the country's high-tech industry. Korea was just on the verge of beating Japan in some industries (Samsung has decimated Sony's lead in consumer electronics in recent years), but Chinese rivals were already plotting their own rise.

For Cisco, however, Songdo was a chance to get in early—not just the steadily evolving market for building automation, which was expected to grow at a tepid 3 percent a year—but the vast new high-growth market for technology-enabled infrastructure: roads, power grids, security, water, and sanitation.[19] The technical challenge of interconnecting disparate sensors, control devices, and number-crunching computers was what Cisco was born of—the company had over three decades of experience weaving the individual pieces of the Internet together. In the beginning, building automation systems were proprietary, so you couldn't mix and match. In the 1990s, several competing standards were developed that allowed devices from different manufacturers to work in concert, but they were far from perfect and for years there was no clear winner. Cisco's vision was to accelerate this integration process and put everything in the city on a "convergence" network, talking to each other using Internet technologies and protocols. If it succeeded, Cisco would reap a nice fee for its hard work and cement itself deep within the basic operations of the city. "The popular technology of our time devotes itself to contriving means to displace autonomous organic forms with ingenious mechanical (controllable! profitable!) substitutes," wrote urban scholar Lewis Mumford in 1961.[20] Cisco seemed poised to write the next chapter in that story.

But for all its promise, it was clear during a visit in the fall of 2009 that *pali pali* urgency was in short supply at Songdo's technology department. From the observation deck of the soon-to-be-completed Northeast Asia Trade Tower—at 1,000 feet above the coast, it is Korea's tallest building—Songdo looks like any of dozens of new towns that have mushroomed on the outskirts of Seoul since the 1980s. Row upon row of identical apartment towers march off to the north and east, bearing oddly Western-sounding luxury brand names like "Hillmark" and "Worldstate." Empty office blocks await the unlucky back-office departments that will be reluctantly relocated from Seoul to the sticks to keep the commercial side of this massive real estate project afloat. Songdo's gambit for foreign investment hasn't worked out as hoped—multinationals simply skipped over Korea to invest directly in mainland China. Pressure was mounting on Cisco and Gale International, the real estate development firm behind Songdo, to fulfill the project's lofty ambitions. In 2011, in a calculated effort to save face, Cisco published a thinly researched white paper frantically touting the social, economic, and environment benefits of smart cities.[21] As Lindsay later explained to me, Songdo had become too big to fail.

From my perch, the "smart" face of Songdo was just as invisible as it was on the ground. A few years later, in 2012, Starbucks and start-up firm Square would announce a retail payment technology that tracks you by smartphone as you enter a shop and lets you pay simply by saying your name. Building a city around RFID cards seems, by comparison, sadly anachronistic. And unlike Digital Media City, an earlier effort to build a small-scale smart city on the edge of Seoul's core, in Songdo the intelligence seemed deliberately tucked behind the scenes. Digital Media City's plans were bold—massive building-sized screens, obelisks projecting social-media streams into public plazas, and free Wi-Fi everywhere. Compared to that design, which echoed Generator in its celebration of the messy human side of the city, Songdo seems intent on engineering serendipity out of the urban equation. In a world of YouTube, FaceBook, and LOLcats, something about Songdo just doesn't feel authentic, fully reflective of our everyday digital existence.

For now, Songdo's potential lies mostly in the somewhat distant future. The real magic of a fully networked and automated city won't be seen until designers start writing code to program truly novel behaviors for entire buildings and neighborhoods. Thinking back to the original problem that faced Paxton as he sketched the Crystal Palace, how could a fully automated city respond to weather automatically as a system, and do it in ways that both reduced the use of energy and created a more delightful, human experience?

Imagine a late summer afternoon in Songdo a few years from now. Instead of thousands of individuals opening shades and adjusting thermostats, the entire city reacts to the setting sun in synchrony. Like desert plants, which open their stomata only at night to minimize water loss, Songdo's smart buildings might order millions of remotely controlled motors to open windows and blinds to catch the evening sea breeze. Air conditioners and lighting are throttled back. Fresh air and the golden rays of the fading sun fill the city's chambers.

This kind of city-scale performance will one day fulfill the potential of building automation. Life in smart cities will be defined by these dynamic, adaptive systems that respond in real time to changing conditions at the very small and very large scale simultaneously. They will fulfill the Frazers' dream of a building that learns from and adapts to us—their moves will be scripted by insights drawn from torrents of sensed data. Indeed, in 2011, speaking at MIT, John Frazer noted that "things that were experimented in at a very small scale in the 1960s and 1970s now can be operated at city scale and even a global scale."[22]

And as smart cities come to know us, they also will come to understand themselves. Deep in the core of Songdo, data centers chock full of CPUs scan the millions upon millions of sensor readings, looking for larger patterns. As this big data accumulates over time, the city's managers will begin to understand its daily rhythms and program new rules about how to direct traffic and power, how to dispatch elevators, how to heat and cool most efficiently and comfortably, and how all of these different actions and movements influence each other. At the very least, they will automate all of the physical systems of the city. At the very best, they will engineer entirely new ways for us to thrive. The infrastructure is being laid, but the ideas and software that will choreograph it will require years, if not decades, of research and development in test beds like Songdo.

Songdo's lackluster technological accomplishments to date aren't its only disappointment. What's been destroyed in this quixotic quest is irreplaceable. Ironically, for a project whose marketers tout it as "one of the world's greenest cities," Songdo's 1,500 acres were manufactured in a massive landfill operation.[23] Where shore birds once nested in ecologically critical coastal wetlands, some 22,500 apartments and over 50 million square feet of commercial space are being built, along with a golf course designed by Jack Nicklaus.[24] "Such green gadgetry seems irrelevant . . ." writes Tim Edelsten, a conservationist based in Korea, "when you realize that a vast natural paradise has been destroyed to create all this new office space."[25]

The Twenty-First Century's First New Industry

Songdo isn't the only smart city on the drawing board. Global urbanization is driving unprecedented investment in cities. Over the coming decades, developing economies such as China, India, and Brazil will spend billions on urban infrastructure to support economic growth and the material needs of a huge new middle class. At the same time, the world's rich countries will have to upgrade existing infrastructure to stay competitive. As new more efficient, more convenient, and more secure designs for infrastructure are crafted, building smart cities will become the first new industry of the twenty-first century.

The price tag for all of those bridges, roads, power plants, water mains, and sewers? An estimated $40 trillion over the next twenty-five years, announced a team of analysts at the consulting firm Booz Allen Hamilton in a 2007 article in the company's magazine merrily titled "Lights! Water! Motion!"[26] Based on the World Bank's 2007 estimate of global GDP of $54.3 trillion, that means slightly less than 3 percent of global GDP needs to be spent on infrastructure each year just to keep up. If anything, the Booz Allen Hamilton analysts' estimate was a conservative tally. Just three years later, in a different forecast for the World Wildlife Foundation, the firm's estimate had ballooned to $249 trillion dollars worldwide from 2005 to 2035.[27] According to a study conducted by Ernst & Young, another consultancy, for the Urban Land Institute, a think tank for the development industry, the United States alone must spend $2 trillion just to repair and rebuild its crumbling networks.[28]

The bulk of this astronomical sum will pay for the old-fashioned cityware of asphalt and steel. That is why South Korea's Posco, one of the world's largest steel manufacturers, is Songdo's main investor. But if even a tiny fraction goes to chips, glass fibers, and software, it will be a windfall for the technology industry. According to Ian Marlow, a consultant who served as the lead technical and business advisor for Songdo's intelligent infrastructure, building-in smart added only 2.9 percent to the project's construction budget.[29] Scale that share planet-wide, and global spending on smart infrastructure is on the order of $100 billion over the next decade alone.[30] That sum spans a big territory, according to one market forecast, including "installing municipal wireless networks, implementing e-government initiatives by providing access to city departments and initiatives through websites, integrating public transportation with intelligent transportation systems, or developing ways to cut their carbon footprints and reduce the amount of recyclables consigned to the trash heap."[31]

Cisco and IBM both have long histories as suppliers to governments, designing systems to bring paper-based bureaucracies into the digital age. Until recently, this was an incremental process that proceeded at the snail's pace of government. The companies' main focus lay elsewhere, on the multinational corporations that were their bread and butter. In 2008 the global recession upended business as usual. The consensus for huge investments in urban infrastructure emerged at almost exactly the same time that governments began planning stimulus programs to buoy underperforming economies. As the private sector choked off spending on new systems practically overnight, an aggressive urgency to push the technologies of global business into government took over.

For these tech giants, the first challenge was making the case for public spending on smart. If you've opened a business magazine or walked through an airport in the last five years, no doubt you've seen the pitch. IBM is estimated to have spent hundreds of millions of dollars alone, educating mayors and concerned citizens about how to upgrade cities. The ads are astonishingly blunt, their claims bold. In the smart city, "Buildings bring down their own energy costs" and "Drivers can see traffic jams before they happen."

The big promise is greater efficiency. For a world facing rapid urban growth, economic collapse, and environmental destruction, IBM and others saw low-hanging fruit in the wasteful ways of government. Technology could fix all that, they argued, by stretching existing resources to deal with the first two problems and ratcheting down the excesses of industrial growth to deal with the third. If we replicated the logistics systems of global business and applied them to the very local problems of cities, it seemed, all would be well. As Colin Harrison, one of the architects of IBM's smart-cities strategy explained it, "For much of the last twenty years, we instrumented the global supply chain. That hasn't happened in city governments."[32]

Remodeling cities in the image of multinational corporations requires three new layers of technology according to Arup, a global engineering giant. The first layer is "instrumentation"—the sensor grids embedded in infrastructure that measure conditions throughout the city, much as companies use GPS trackers, bar codes, and cash-register receipts to measure what is going on in their businesses. This raw data is fed into "urban informatics" systems that combine data-crunching hardware and software to process the signals into usable intelligence and let us visualize and discover patterns that can help us make better decisions. Finally, an "urban information architecture" provides a set of management practices and business processes to tell people how to use the results of these computations to get their work done and cut through red tape and bureaucratic barriers. As the company argued in a 2010 white paper, "the smart city is so different in essence to the 20th century city that the governance models and organisational frameworks themselves must evolve."[33] Together, these three layers will allow us to rewire governments by design, transforming the way they work internally and together with outside partners and citizens.

To understand how all of this might help cities, look at the effect of technology on air transportation over the last few decades. For customers, interactions with airlines often have a Kafkaesque tenor of confusion and disdain. But behind the scenes, an arsenal of sensors, informatics, and information-driven business processes are at work, coordinating the movements of millions of passengers, crew, baggage, and planes. It was estimated in the late 1990s that "50,000 electronic exchanges of all sorts" were required to get a single Boeing 747 off the ground, from booking seats to ordering food and fuel.[34] In today's highly instrumented and networked air transport network, millions of digital transactions orchestrate each flight. Shared through global networks, these data guide the decisions of dispatchers, travel agents, and passengers in real-time. Innovations like dynamic ticket pricing, automatic rebooking, and mobile flight status alerts all ride on top of these systems. While it rarely feels so, the air transportation system is among the smartest infrastructures in our cities.

It's a tough pitch to resist. For a world that seems increasingly out of kilter, rewiring cities with business technology is a seductive vision of how we can build our way back to balance. As cities struggle to grow while simultaneously improving public services and reducing carbon emissions, something has to give. If a modest investment in smart technology can deliver greater efficiency, it will pay for itself—a rounding error, really, on the staggering infrastructure investments needed.

Nowhere is the need more clear than in our aging, obsolete, and inadequate electric power grid.

Power Platform

We take few things for granted more than the ubiquity of electrical power in modern cities. We are only conscious of its existence when it fails. And while a surprising number of us still avoid the Internet, only a handful of sects shun the many conveniences of electricity. In 2008 the world's power plants produced 19.1 trillion kilowatt-hours of electricity, and global generating capacity is expected to nearly double by 2035.[35] This growth will be driven by urbanization, as businesses in developing countries build new factories and workers spend their newfound wealth on electrical appliances. Huge new sources of demand will also come online as urban infrastructures that have traditionally been powered by fossil fuels shift to electricity. Electric cars and buses will refuel from the power grid instead of gas stations. Geothermal heat pumps, which use the steady temperature of the earth's crust to efficiently heat and cool homes and buildings, will replace oil and natural gas–fueled boilers.

The grid will get bigger, but also more complex. Adding renewable sources of energy like solar panels and wind turbines to the grid dramatically increases the need to move electricity around like Internet packets. The sun doesn't shine evenly and the wind shifts, creating a fickle flow of power that needs to be balanced across regions and over time. Add to that our own variable demands for power and the challenge gets very complicated very quickly.

As you read this book, chances are you are being lit, cooled, or transported by something invented by General Electric or Siemens. Long before Cisco jumped into city building, these companies laid the lines that, to borrow from GE's 1980s marketing campaign, "bring good things to life." But perhaps GE was too modest. They don't just bring good things to life. They make modern life possible. The scale of these companies is breathtaking. Both employ hundreds of thousands of people and generate more than $100 billion in annual revenues. But it means they are well matched to the enormous engineering challenge of providing mobility, sanitation, energy, and communications for seven or eight billion middle-class city-dwellers worldwide by this century's end.

Their first task in this century is to rebuild the electric power grid they built in the last one. Overhauling the power grid is an urgent priority for smart cities because without a stable supply of electricity everything comes to a stop. When a tsunami struck Japan in 2011, triggering the shutdown of most of the nation's nuclear generators, the multistory digital screens of Tokyo's Shibuya Crossing—the Asian equivalent of Manhattan's Times Square—went dark for weeks. Normally crisscrossed by mobile phone–toting "smart mobs," as author Howard Rheingold dubbed them, it is a place that lives in my memory as the paragon of future urbanism. Tokyo survived its digital lobotomy—there's still enough of the conventional infrastructure in place to live life manually, so to speak. But in future cities even the most mundane tasks will draw upon sensors, computers, and communications networks scattered across the cloud. Electricity, even more than the digital data it conveys, will be the lifeblood of smart cities.

Rewiring the world's power grids is a massive undertaking. Siemens constructed the first public electric utility to power a network of forty-one streetlamps in the London suburb of Godalming some 130 years ago. In that short time, we've built up a massive complex of wires, transformers, and power plants that stretches across the globe. Massoud Amin of the University of Minnesota argues that "the North American power network may realistically be considered to be the largest and most complex machine in the world." In a 2004 inventory, he counted more than fifteen thousand generators in ten thousand power plants hooked up to hundreds of thousands of miles of distribution lines, representing nearly one trillion dollars' worth of public and private investment.[36] In the United States alone, power producers booked some $368 billion in gross revenues in 2010.[37]

The power grid and phone system were both born during the great urban boom of the late nineteenth century. While the phone network's guts were upgraded several times in the twentieth century—machines supplanted human operators, fiber-optics replaced copper cables—the power grid seems stuck in time.

But why did the phone network evolve and the power grid stagnate?

AT&T's monopoly in telecommunications, built by industrialist Theodore Vail in the early 1900s with financial backing from J. P. Morgan, sacrificed innovation for expansion and consolidation.[38] But compared to the scant investments in research by the electric power industry, which is a highly fragmented patchwork of thousands of privately owned and municipal utilities and rural cooperatives in the United States and Canada, it was a veritable renaissance of invention.[39] By the 1970s, there were enough breakthroughs on the horizon—fiber-optics, cellular telephony, and digital switching were all introduced during that decade—that investors began pushing aggressively for deregulation. In 1969, the newly formed MCI had won government approval to deploy a wireless trunk across the Midwest, connecting Chicago and St. Louis in a daisy chain of towers linked by focused beams of microwave energy. Directly competing with AT&T's long-distance business, MCI's arrival was a watershed, launching a decades-long era of innovation in telecommunications infrastructure around the world. This massive, sustained investment in research, development, and construction is the reason that today, instead of waiting years for a government-owned phone company to grant you a line, in almost any country you can walk into a store, buy a mobile phone, and be instantly connected.

Digitization of the phone system in the 1980s accelerated the pace of change. Early telephone networks required all calls to be manually completed by a human operator, who would use patch cables to cross-connect lines to close a circuit. In 1889, Kansas City undertaker Almon B. Strowger invented an electromechanical device to automatically switch calls, motivated by his belief that telephone operators were diverting incoming calls to his competitors.[40] A century later, the introduction of digital switches in the 1980s turned voices into data. This allowed more calls to be squeezed onto the same trunk lines. More importantly, it put intelligence in the network. Creating new services like call waiting, voice mail, and caller ID was simply a matter of writing new switching software. Operators could also see and direct the flow of calls in real time, at any point in the network.

Digitization proved such a versatile platform for innovation that it allowed the phone network to spawn the Internet that would eventually eat it. When I arrived fresh out of college to work at AT&T in 1996, the Internet was a small trickle of traffic inside the company's national frame-relay network. Originally built to shuttle voice calls around the country, AT&T's grid was capable of carrying many other kinds of data too—financial transactions and Internet packets as well. As part of an elite tech-support team for the company's brand new Worldnet dial-up Internet service, I fielded some of the most difficult calls, helping AT&T executives figure out how to dial home from a business trip in Singapore, for instance. In the evenings, as the nation came online in droves, I'd gaze up at the big control board, watching as the network shunted traffic around choke points. On the rare occasion that the system's self-healing stopgaps failed, a few keystrokes could reroute transcontinental traffic through Kansas City instead of Chicago. Less than twenty years later, the balance of traffic on the world's networks has flipped—most voice calls are now transmitted by Internet Protocol.

Back in the world of electric power, you can forget about tracking electricity, much less directing it. To be fair, physics has stacked the deck against the power grid. Big flows of electricity can't be chopped up and piped about the way digital bits can. Digital telecommunications networks use temporary containers called buffers to manage congestion at choke points. But keeping the power grid running smoothly is more of a balancing act than a job of directing traffic. Storage for electrical grids is much more expensive and problematic—instead of RAM chips, utilities must install massive flywheels, batteries, and capacitors to throttle the flow of power. Adding to the challenge is a lack of instrumentation. Unlike digital telecommunications networks, which by design are fitted with all kinds of flow sensors, the power grid is dumb. In Arbon, the Swiss town that Siemens has chosen as a guinea pig for its smart grid technology, the power company's director readily admits that "even today, neither consumers nor suppliers know exactly when electricity is flowing through power lines, or how much of it is flowing."[41]

What's perhaps more shocking is the age of the power grid, and how much of it is undocumented. Utility companies don't know exactly where a lot of the infrastructure *is*. After the September 11 attacks in New York, I often rambled through the streets of Lower Manhattan late at night. Peering down into excavation pits, I watched as crews from Con Edison scratched their heads in bewilderment, struggling to untangle a century's worth of cables unearthed in some subterranean vault. This is an extreme case, but most of the North

American power grid dates from the 1960s. According to the head of the International Brotherhood of Electrical Workers, the electricians' union, the average age of transformers (electrical devices that change the voltage of flowing current) in service in 2007 was forty years, which also happens to be their useful working lifetime.[42] As the editor of trade rag *EnergyBiz* put it, "We are talking about equipment deployed before a man walked on the Moon, before cell phones and the Internet, when Frank Sinatra was in his prime."[43]

Siemens took a few more years than IBM or Cisco to refocus its ambitions on smart cities, in part because it's a much larger company. Big ships are harder to turn. But in 2011 it made a massive shift by reorganizing more than 85,000 employees into a new Infrastructure & Cities division. Building smart cities is in fact a return to the company's roots. Unlike GE, which was founded as an electricity company, Siemens actually got its start building communications networks. The first Siemens company, Telegraphen-Bauanstalt von Siemens & Halske, strung Germany's first inter-city telegraph line between Berlin and Frankfurt in 1848.[44] Since then, the firm has long dominated infrastructure markets that depend on electricity—not just power grids but also electric trains, an industry it leads to this day.

While Siemens still builds smart systems for telecommunications and transportation, the smart grid plays a special role in its vision for cities because, writes Jeff St. John on the *GigaOM* blog, it's "one of the few corporations out there that can lay claim to almost every share of the world's current grid infrastructure, building everything from gas and wind turbines to high-voltage transmission cables to sensors and controls that monitor and manage the delivery of power to homes and businesses."[45] Targeting nearly $8.5 billion (€6 billion) in annual smart grid business by 2014, CEO Peter Löscher boasted, "We're on the threshold of a new electric age."[46]

As consumers, we think of the smart grid mostly through our growing experience with smart meters. Smart meters are to your old electric meter what a smartphone is to your grandmother's Bakelite

1950s rotary phone. It's a souped-up, networked upgrade that constantly reports back to the electric company a stream of data about your power consumption, including when it detects blackouts and brownouts. The more advanced models can manage power-hungry appliances in your home. In-Stat, a market research firm, projects that by 2016 fully three-quarters of American electric meters will have been converted to smart meters.[47] While these are the most visible endpoints of the emerging new grid, Siemens actually sold off its smart-meter business a decade ago. Its true ambition is to become a Cisco for electricity, providing the brains inside the smart grid, the software and switches that manage the behind-the-scenes balancing act that keeps the juice flowing.

The power grid shell game isn't only about keeping the lights on, but doing it cost-effectively while letting loose as few emissions as possible. What makes this process hard is the erratic demand for electricity, particularly in cities. Electric utilities deal with irregular ebb and flow by building two different sets of power plants. Base-load plants serve the minimum demand for electricity that stays constant year-round. These highly efficient plants can be run more or less continuously at near-full capacity. But because demand for electricity in a place like New York can spike as much as 40 percent on the hottest summer afternoons, utilities also build "peaking" plants that can be quickly brought online as needed. While peaking plants can also be highly efficient—most are natural-gas–powered turbines—they are far more costly per unit of power to build and run. If only the peaks could be evened out, fewer peaking plants would be needed and utilities could focus more on ruthlessly fine-tuning base load plants to be as lean and clean as possible.[48]

Smart grids offer two tricks to even out the peaks: load shifting and load shedding.

Load shifting, the gentler of the two, tries to spread demand for power away from peak periods of demand through price incentives. In their simplest form, smart meters allow businesses and consumers to see the true cost of generating electricity during periods of high demand. As they fire up those costly peaking plants, utilities simply pass the higher generating cost along to consumers. Dynamic pricing can dramatically reduce swings in demand for power and increase overall generating efficiency, but load shifting can also be automated and proactive. Smart meters that communicate directly with smart appliances might automatically reschedule a load of wash for later in the day when demand and prices are likely to fall.

Even the most sophisticated load-shifting scheme will one day meet its limit. That's when utilities wield their trump card—load shedding—a kind of targeted blackout. Traditionally, load shedding was a manual process. Utilities would cut deals with large users of electricity like factories and universities to shut down power during peaking crises in return for a discount on their regular rates. Smart meters will allow these miniblackouts to be replaced by sophisticated surgical drawdowns on sacrificial facilities and equipment. A university might agree to have its dormitories or office lighting shut off while service to sensitive laboratory instruments, for instance, is maintained. A factory could shut down a production line in stages to reduce the need to discard unfinished products damaged by idling.

Without smart controls like these, the grid's problems will worsen rapidly. Even as demand surges, building new power plants only gets harder as NIMBY-led resistance to plant construction spreads in many countries. The wiggle room that once existed in the form of reserve generating capacity is fast disappearing, raising the possibility of regular blackouts in the future. During the 1990s, demand for electricity grew by 35 percent in the United States, but generating capacity increased by only 18 percent.[49]

According to Siemens, smart grids will help utility engineers sleep at night, since load shedding and load shifting could reduce national electricity needs by up to 10 percent.[50] Environmentalists will cheer because improved demand management removes a key obstacle to greater reliance on renewable generating sources, which are notoriously unreliable base capacity—the sun doesn't always shine and the wind doesn't always blow. Even hydropower generated at dams depends on reliable seasonal rains to fill up rivers. Greater ability to reduce demand when the supply of green power falters will reduce the need for fossil-fuel powered backup plants.

But beyond just keeping the lights on, the smart grid could finally unleash the kind of innovation in energy services that we've become accustomed to in telecommunications. Start-up firms could audit and manage our home's electricity use in return for a small cut of the savings off our energy bill. In a world where Siemens forecasts that electricity prices could change as often as every fifteen minutes, we'll be relieved to have a piece of tracking software automate the process.[51]

By allowing us to account for all of the power we put in and take out of the system, the smart grid will also allow us to add a social layer to the production, distribution, and consumption of electricity. Imagine connecting your smart meter to Facebook. You might dare your neighbors to cut back as much as you do, in a game to save the earth played out on the smart grid of your neighborhood. Or, as Eric Paulos of the University of California, Berkeley, proposes, we can decommodify energy by creating sensors to document how, where, and by whom it was generated and making this information available during transactions. "Is it fresh energy? Is it local energy?" he asks. What if instead of sending a text message, a child could send mommy the 100 watts she just produced on a power-scavenging swing set?[52] Scale this model up, and it is possible to imagine a rich trade of power between many producers and consumers, incentivized by any number of causes, interests, or goals. A social meta-layer on the smart power grid could have enormous impact on our consumption choices.

Deregulation now allows many consumers to choose which producer to buy their electricity from, even as that power is still delivered across a single grid controlled by the local utility. Power providers compete on price and carbon footprint. But we are moving into a world where the data about electricity will become as valuable as the power itself. Already, start-ups like Arlington, Virginia–based Opower are showing how smart meters will enable utilities to bundle information and services with basic electricity to add value. These tools can help consumers save money, and are very convenient. They also hold the potential to make us more understanding and conscientious about how we use electricity. Choosing your power provider in the age of the smart grid will be more like choosing a mobile phone carrier is today. The grid itself is a commodity. All the value is in the add-ons.

The Fourth Utility

The power grid is the circulatory system that delivers the lifeblood of electricity throughout cities. Data networks are their nervous systems, shuttling messages to and fro. Much as we are upgrading the power grid, new communications networks are upgrades to systems first built during the rapid growth of cities in the nineteenth century. In fact, the first urban digital communications network was the telegraph. The dots and dashes of Morse code were as binary as the 0s and 1s of the digital computer.

The telegraph didn't appear out of nowhere. It was invented specifically to meet the growing need to coordinate vast commercial and government enterprises. By the mid-1800s, the industrial revolution was hitting full stride. Steam-powered machines allowed businesses to make and transport goods on such a massive and rapid scale that human managers couldn't keep up. It was a full-fledged "crisis of control," as sociologist James Beniger described it. "Never before had the processing of material flows threatened to exceed, in both volume and speed, the capacity of technology to contain them."[53] Throughout the first half of the 1800s, tinkerers in Europe and the United States worked feverishly to develop systems for transmitting messages via wire using electrical pulses. The race culminated in the 1840 patent for the Morse-Vail system. Telegraph systems fueled the expansion of intercity trade by synchronizing railroad operations. For the first time, business information could move faster than the speed of travel.

Much like today's new communications technologies, the telegraph inspired its own set of urban visions. In the 1850s, just as Siemens was stringing telegraph lines between German cities, the Spanish city of Barcelona broke free from the shackles of history and began to expand and modernize. Hemmed in for centuries by its city wall, rapid industrialization had turned the city into one of Europe's most densely populated. In 1854, authorized by royal decree, citizens eagerly began tearing the wall down by hand. As one historian recounted the riotous affair:

> As soon as the news of the government's long-desired permission to pull down the wall was known, there was a general rejoicing in the city, and its shops were emptied of pickaxes and crowbars overnight. Almost every citizen rushed to the wall to participate in its demolition, either by using the appropriate tools or by supporting orally those who were actually doing the work. The wall was, probably, the most hated construction of that time in a European city. . . . It took twelve years to pull them down, which is not a long time when we remember that they had stood erect for nearly one and a half centuries.[54]

The way was clear for the city to modernize and grow by exploiting the new technologies of the "control revolution," as Beniger dubbed this great period of technological and organizational transformation.

Outside the walls lay a blank canvas of sparsely settled countryside onto which Ildefons Cerdà, a visionary civil engineer, laid out a new district designed around the potential of the railroad and the telegraph. In his 1867 opus *Teoría General de la Urbanización* (*General Theory of Urbanization*), Cerdà expressed his fascination with these new technologies, contrasting the "calm and tranquil, almost motionless man of the earlier generations that preceded us" with the "active, daring, entrepreneurial man . . . who in just minutes transmits and circulates his news, his instructions, his commands right around the globe."[55] His plan for L'Eixample (literally, "the extension"), embraced these new technological capabilities.

Cerdà didn't just dream. His sketches provided precise diagrams for accommodating the telegraph. "It is indispensable for the underground Extension works to include a way to accommodate this service in the most convenient and economic way possible . . . ," he wrote, "for this it is only necessary to leave enough room in the ducts for the wires to be laid."[56] His plan called for "four longitudinal conduits for each street: 1) For the distribution of drinking water. 2) For the disposal of sewage. 3) For the distribution of gas. 4) For the laying of telegraph wires."[57] In Cerdà's vision the telegraph would be a fourth utility for the industrial city, a network that author Tom Standage has called "The Victorian Internet."[58]

Over 150 years later, Cisco Systems has unwittingly commandeered Cerda's schema as it plans the next generation of telecommunications networks for cities throughout the world. "Visionary countries . . . understand that the network is the fourth utility," proclaims the company's chief globalization officer, Wim Elfrink, "enhancing global competitiveness, innovation and standard of living."[59]

Today Cisco is becoming a household name, but few people realize the company is an industrial giant on the scale of Chrysler or Dow Chemical, with some $40 billion-plus in annual revenues. Founded in 1984 by husband and wife Len Bosack and Sandy Lerner, who built Stanford University's campus network in the early 1980s, Cisco has grown into the world's leading supplier for the sophisticated switches and routers that power the Internet. Cisco's products not only push bits around offices, schools, and homes, but also sling them back and forth across undersea cables that link continents. It's one of Silicon Valley's largest and most-watched bellwethers. For a brief period in March 2000, at the height of the telecom bubble, it was the most valuable company in the world.

But with size comes stagnation. Finding growth opportunities has become a constant struggle for Cisco, and to make a dent on the bottom line it needs to have billion-dollar payouts. The company's ambition to become the new plumber of smart cities isn't limited to Songdo, or even all of the to-be-built pop-up cities of Asia. The firm wants to control the nervous system of the entire urban world.

Injecting smart features into existing cities is a daunting prospect. Just making a single building smart is a monumental task of interconnection and translation. They are riddled with special-purpose networks built out in recent decades that can't talk to each other. A single building might have one set of control wires for elevators, another for heating and ventilation, another for security, and yet another for lighting. Integrating a whole city full of these legacy networks presents an almost intractable problem.

To Cisco, however, the problems that hamper would-be smart cities look a lot like the ones that universities and corporations faced in the early days of the Internet. The challenge then was connecting hundreds or thousands of independent local area computer networks (LANs) into an integrated Internet. The challenge now is figuring out how to interconnect fragmented city infrastructure by using the Internet to bridge these gaps in the urban fabric. Soon after signing up as Songdo's chief technology supplier, Cisco spun up a smart city engineering group at its new "second headquarters," the Globalisation Centre East in Bangalore, India.[60] "Today, urban centers struggle with hundreds of different systems and protocols that do not interoperate," a brochure touting the new lab proclaimed. "If these systems converge onto a single open-systems based network, significant opportunities for productivity, growth, and innovation can be unleashed."[61] It was a compelling if somewhat quixotic vision of progress for a fast-changing urban world.

As a corporate strategy, it seemed like a slam dunk. Cisco's network would unscramble the Tower of Babel that is our urban infrastructure. The company would extend its long-held dominance as the Internet's traffic cop to the networks that connect buildings, vehicles, and urban infrastructure to city-scale control systems. Interconnection would enable new city-scale applications and drive growth in data traffic. And every extra bit traversing a neighborhood was another bit for Cisco's high-profit-margin routers and switches to direct. The fourth utility, deployed to interconnect the physical world, promised to be at least as big an opportunity as the original Internet, which was built to interconnect virtual worlds.

But just as the market for smart city networks was shaping up, video flooded onto the Web via fiber-optic networks laid during the telecom boom of the 2000s. Integration and automation of building and infrastructure systems might provide steady business for decades to come, but the rise of video communications held out the possibility of a wild, bucking bull that Cisco could ride to astronomical heights of profitability. Since the earliest days of television, the videophone had been one of those inventions that was perennially just around the corner. Decade after decade, prototype after prototype had failed to capture the public imagination. Finally, it seemed, the world was ready for faces to accompany voices coming over a wire.

Almost overnight Cisco's entire smart-city pitch shifted to video. In 2011 it released a "Visual Networking Index" that highlighted the coming crush. By "2015, the gigabyte equivalent of all movies ever made will cross global IP networks every 5 minutes," the company predicted.[62] But instead of quenching the fire, Cisco was throwing on fuel. Its multiscreen, high-definition TelePresence videoconferencing systems were selling very well, for hundreds of thousands dollars per unit. Beginning in 2006, it began to experiment on itself to build a business case for the technology, deploying over 250 units in 123 cities worldwide. In 2008 the company announced it had saved $90 million by eliminating travel for nearly 17,500 face-to-face meetings.[63] In 2010, it acquired Norway-based Tandberg, a manufacturer of desktop videophones, and cut a deal to install the units in apartments throughout Songdo's residential quarter.

Just as it was ramping up production of TelePresence, Cisco was putting its own spin on Songdo's significance for a rapidly urbanizing China.

■ ■ ■

"Of course I can see you! You're as big as a wall!" exclaims a venerable Chinese gentleman. In a luminous, light-filled apartment in the Shanghai of 2020, we snoop on a joyous video call between an elderly couple and their friend, discussing the upcoming evening's reunion for a wedding anniversary celebration. A cinema-sized display occupies the entire wall of their living room.

Shanghai's Expo 2010 was arguably the most important international showcase since the 1939 World's Fair in New York. And much like that earlier exposition, a phalanx of corporations looking to cash in on the next building boom promulgated visions of how to shape the landscape of a newly prosperous nation. The theme was, simply "Better City, Better Life." In 1939, General Motors' exhibit envisioned how one technology, the automobile, could power a future migration of Americans out of cities into the suburbs. But in Shanghai, Cisco's pavilion demonstrated how a very different technology, high-definition videoconferencing, could restore harmony to a China fractured by a massive migration from the countryside *into* cities that was just reaching its climax. The million-plus families displaced in the reconstruction of Shanghai as a modern, global city would be stitched back together by the Internet.[64]

The heart of Cisco's show was a seven-minute video depicting a day in the life of 2020 Shanghai.[65] Even before we meet the elated senior citizens, the film opens in the city's control center, where a fast-approaching typhoon has just been detected by an advanced weather-tracking computer. As capable government managers calmly order emergency preparations, the story abruptly cuts to domestic life. We see the lives of two young couples unfold on the screen. One is on the verge of a breakup, the other about to have a baby. High-definition video communications propel the events. As her first contractions begin, the expectant mother consults her doctor from the kitchen counter, then summons her husband in his car halfway across the city. Intuitive, mobile, and effortless, high-definition video keeps the city's residents in near-lifelike contact at a distance and on the go.

Cisco's vision painted the aspirations and fears of modern China with coarse strokes. It promised to recapture all that had been lost in the country's rapid urbanization, which in two decades had transformed the Chinese family more fundamentally than the two millennia that preceded them. Traditionally, Chinese lived in multigenerational households, with many members of extended families under one roof. But the move to cities brought a shift toward more Western-style nuclear households of just parents and children. In Cisco's future Shanghai, orphaned elders would become the early adopters of video chat.

As the typhoon closes in, the characters move through an increasingly threat-filled city. But as in a Greek myth, the heroes in Cisco's vision of Shanghai in 2020 do not act entirely of their own free will. Like the gods of Mount Olympus, city managers peer into a miniature holographic simulacrum of the city and its inhabitants. Instead of atmospheric clouds, their aerie rests in a computational cloud. Their omniscience comes not from divinity but from a massive grid of sensors that can seemingly track anything—rainfall, traffic jams, even the movement of individual citizens. By remote control of infrastructure and instantaneous dispatch of responders, they possess an omnipotence that no mayor has ever known. Above all, order is maintained in this patently paternalistic view of the future. Shanghai's residents of 2020 have surrendered to the guardians behind the screens.[66]

It is a provocative vision, this city of screens. For China, surely, but for the rest of us as well. In America it could mean rewiring our sprawling suburbs, saving energy and reducing traffic by replacing car trips with video calls. If this future catches on, hooking up cities for mass video communications could power Cisco's profits for years to come. It's a well-worn cliché that the only people who get rich in a gold rush are the ones selling picks and shovels. But beyond just peddling tools and equipment, if Cisco's network becomes a true "fourth utility," all bets are off.

Hints of the potential are emerging in Songdo, where the company will install ten thousand TelePresence screens in homes, offices, and schools by 2018. The screens come included with new apartments, and unlimited video calls will cost just $10 per month. But Songdo U.Life—a new joint venture between Cisco, the developer Gale International, and Korean tech giant LG—will also launch a kind of app store, where residents can subscribe to a whole host of new interactive video. As Eliza Strickland reported in *IEEE Spectrum*, "a resident could start her day with a live yoga class; later her child could get one-on-one English lessons from a teacher across the world."[67] Much like Apple's App Store, U.Life and Cisco will exact a healthy vigorish from service providers who want to plug in to its hi-def grid.

Over the last decade, Cisco's fortunes have whipsawed between growth and collapse, first riding the telecom bubble of the late 1990s to near-oblivion in 2000, and then slowly tracking the broadband expansion of the next decade back to stability. Today, facing a future of intense competition from China's Huawei, Cisco is taking the boldest bet on smart cities of any technology giant. Alone among them, it challenges us to radically rethink how we build and live in them. One of the company's ads in *The Economist* magazine that featured the skyline of Beijing imprudently asks, "Is this really the end of cities as we know them?" The answer, a punt: "Check back in 20 years."

Untethered

For the last thirty years, the Internet has been a thing that we "dial up" to or "jack into." While cyberspace was an ethereal place, the process of getting there meant making a very real and direct physical connection.[68] That's no longer the case. We've untethered ourselves from the Internet's wired backbone: our dealings with it now are almost exclusively via radio waves.

The networks that make our mobile connected lives possible are the newest and most crucial infrastructure that will power smart cities. Yet, possibly because they are mostly invisible, we can't seem to figure out what to call them. None of the commonly used monikers quite capture their importance. One can only wonder how long the oddly durable anachronism "wireless" will stick around. "Cellular" (and the even worse "cellular telephony") is a technician's term, mostly confined to use in the United States, which describes the network's underlying architecture of towers. It's like calling the Internet "distributed packet-switched computer networking" instead of the "Web." "Mobile" starts to get at the essence of why people find these technologies so utterly appealing but misses one big aspect of how we use them. Most of the time we aren't moving, we're sitting still.

There is a more fitting adjective that captures both the technology and what it is doing to us. In the 1990s, as the US military contemplated battlefield communications in the future, it adopted the term "untethered." The idea is apropos. Roaming across the room or across the city, we are, in every sense, free of the cables that once tied us to our desktop. It's hard to think of a technological revolution that has snuck up on us with such little fanfare. Perhaps that's because it has been such a long, slow process, moving forward in glacial steps throughout the twentieth century as ways of organizing society and structuring human settlements have evolved.

Mobile radios are now nearly a century old. In 1920, radio enthusiast W. W. Macfarlane demonstrated a setup for two-way communications from a moving vehicle in the Philadelphia suburb of Elkins Park. As *Smithsonian Magazine* recounts it, "With a chauffeur driving him as he sat in the back seat of his moving car he amazed a reporter from *The Electrical Experimenter* magazine by talking to Mrs. Macfarlane, who sat in their garage 500 yards down the road."[69] The horrors of World War I's trench warfare no doubt in his mind, Macfarlane immediately saw the value of his invention for a mobile military. In a prescient prediction of our modern, networked infantry, he envisioned how "A whole regiment equipped with the telephone receivers, with only their rifles as aerials, could advance a mile and each would be instantly in touch with the commanding officer. No runners would be needed."[70] The Second World War would prove Macfarlane right. By 1940 engineers at Motorola had perfected a rugged mobile FM radio transceiver that could be carried in a soldier's backpack. The original "walkie talkie," Motorola's SCR-300, weighed just thirty-five pounds, and with a ten-mile range was often the only line of communication between field commanders and fast-moving units on the front line.[71]

American servicemen returned home with a deep appreciation for the advantages of mobile communications in combat, and an eagerness to turn this novel technology to commercial purposes. AT&T launched the first US mobile phone network in Saint Louis in 1946 with a single call from a driver in his car. The system was based on technology developed for police use during the preceding decades. In 1928 the Detroit Police Department installed wireless receivers in cruisers, creating the first radio police dispatch system. A simple one-way broadcast, station KOP played music in between official announcements to comply with its federal licensing as an entertainment station (there were no official law-enforcement radio bands at the time).[72] By 1933 two-way radios were developed and quickly deployed nationwide after successful testing by police in Bayonne, New Jersey.[73]

With just a single transmitter for receiving calls, and a handful for the return signals, the primitive radiotelephone system launched in 1946 could handle only three simultaneous calls across an entire city in a party-line arrangement—you had to listen for a clear channel before making a call. By 1948 service had been expanded to over a hundred cities, but with only five thousand subscribers nationwide, it remained a costly luxury for the rich and powerful. An upgrade in 1965 increased capacity to forty thousand subscribers and allowed customers to dial directly rather than use an operator. But scarcity still reigned, and service was rationed by state regulators. Some two thousand subscribers in New York squeezed into just twelve shared channels. The average wait time to make a call was thirty minutes.[74]

Constrained by the need to share airwaves, the mobile telephone's future seemed limited to a niche. But there was another way to expand; a clever scheme for a high-capacity mobile phone system had moldered in a file cabinet at Bell Labs, AT&T's research center, since 1947.[75] Instead of using a single transmitter, cities could be divided into a mosaic map of hexagonal zones or "cells." The precious channels could then be reused in nonadjacent cells without fear of interference. Driving from one side of the city to another, a phone might hop on and off the same frequencies several times. Some fancy engineering was needed to coordinate the handoff between towers, but by the late 1970s new digital switching capabilities in the public telephone network had given the grid enough smarts to handle it. "Cellular telephony," the awkward moniker loved only by the engineers who coined it, was born. Every time you see a mothlike tangle of wireless antennas sprouting on the roof of a building, that's the hub of a cell of wireless callers moving through the surrounding area. From that point calls are routed over a "backhaul" wire into the region's landline grid. As communications scholar George Calhoun puts it in *Digital Cellular Radio*, the cellular network "is not so much a new technology as a new idea for organizing existing technology on a larger scale."[76]

Breaking the wireless network up into cells had the added benefit of reducing the amount of power needed for phones to talk to the tower. Rather than send a signal to a tower a dozen miles away, your phone would talk to an antenna just down the street. Less power per call meant smaller batteries, paving the way for much more portable devices. The brick-sized Motorola phones of the 1980s, though they seem immense to us now, were at the time a huge breakthrough in portability and convenience.

The first generation of cellular networks improved capacity by an order of magnitude over the earlier radiotelephone system—from tens of thousands to hundreds of thousands of subscribers. Prices fell rapidly too, as regulators introduced competitive licensing for different frequency bands, further stimulating demand. But once again, the density of demand in cities pushed the system to its breaking point. On Wall Street, in Hollywood, and inside the Beltway, the nation's business and political elite, with their incessant chattering, quickly exhausted the new capacity. And so, in the late 1980s, having already sliced up the city geographically, engineers began slicing the airwaves in time.

First-generation cellular networks, which you may recall as "analog" cellular, worked like the old Bell telephone system. When you dialed, you took over an entire channel for the full duration of your call. Second-generation cellular networks, rolled out in the early 1990s, used digital signaling, which only took up a channel when you were actually talking. When there was nothing being said, part of someone else's call could be smartly shoved into the gaps in transmission. A channel that once carried a single analog call could now carry six or more calls. Digital signal processing brought other benefits—it eliminated the echoes, static, and interference that plagued analog networks and employed strong encryption to put an end to illicit snooping; again it required less power to transmit, further shrinking battery bulk.

Of course it still wasn't enough. Demand kept growing, as millions—entire city populations—could untether. On top of voice traffic, data traffic from wireless e-mail, web browsing, and media uploads and downloads exploded. A third generation ("3G") of infrastructure with more frequencies, and more advanced compression schemes that squeezed more bandwidth out of them, were launched. Engineers took out their scalpels and sliced up existing cells into ever-smaller "microcells" and "picocells" so that the same spectrum could be reused hundreds or even thousands of times across a city.

Despite its slow and often painful evolution over the last century, our untethered infrastructure's greatest challenges lay ahead. The unexpected success of smartphones and tablet computers has placed huge strains on carriers' data networks, as they suck down screenfuls of data from the web. The launch of the iPhone in 2007 overwhelmed the feeble cellular networks in cities with dense clusters of early adopters like New York and San Francisco. Since then, global mobile data traffic has doubled every year.[77]

Video communications may be the killer app for smartphones, but they are also killing the networks, which may be unable to keep up with demand. As 3G networks are upgraded to even faster 4G specs, streaming video to a high-resolution device like the iPad 3 can burn through a subscriber's monthly data allowance in just a few hours.[78] Ericsson, a maker of both cellular handsets and network equipment, reported in 2011 that "the top 5 to 10 percent of smartphone users are willing to spend up to 40 minutes a day watching online video."[79] As a result, AT&T projects that its network will carry more data in the first two months of 2015 than in all of 2010. By then, wireless carriers could be spending over $300 billion annually to satisfy our thirst for bandwidth (not including the actual cost of building the networks), a sevenfold increase over 2010.[80] This assumes they can obtain the needed frequencies—with the concentration of such high-bandwidth users in dense cities, it may be physically impossible for wireless carriers to keep up. "If you had a quarter of the population of Manhattan watching a video over their handset," explains telecom policy scholar Eli Noam, "it would take approximately 100,000 cell sites, or a huge amount of additional spectrum."[81]

Another potential black swan for our untethered grid is the Internet of Things. As of yet, there are few killer apps for connected things that could compete with video as a source of data traffic. But wireless will be a natural medium for connecting itinerant things to the cloud, for the same reasons it appeals to people. Even for stationary things, hooking into a wireless network is now faster, easier, and cheaper than stringing a wire. When New York City wanted to deploy a real-time traffic control system in 2011, it didn't string fiber-optic cables to all twelve thousand–plus traffic lights.[82] Instead, it simply piggybacked an uplink to its half-billion-dollar public safety wireless net, NYCWiN.

The future of mobile networks isn't all doom and gloom. Up until now, every time wireless data speeds have taken a step forward, there's been a new bandwidth-hungry app incubated in the world of desktop computers ready to overwhelm them. The fact that light waves traversing a fiber can carry far more information than radio waves in the air has meant there's always a huge speed gap between the two media. But as we move into a world where wired connections are a thing of the past, and instead of having two classes of broadband, we may only have one, will that drive innovation in services that can live within the more restrictive bandwidth diet of wireless networks? The evolution of mobile apps, which deliver huge value even while volleying relatively fewer bits back and forth to the cloud, seems to point towards that scenario. Or will some new scheme to expand the capacity of untethered networks break this historical pattern of scarcity?

As uncertain as the future for our public untethered networks is, new investment is likely to help ease the crunch. According to IDC, a market research firm, the cellular industry could be spending as much as $50 billion annually by 2015.[83] Governments are moving to free up more spectrum by reallocating bands abandoned by television broadcasters. Still, we are reaching limits on how much smaller cells can get. In big, dense cities, cell sites are often only a few hundred feet apart.[84] At that scale cellular networks will begin to blur with the vast but fragmented constellation of Wi-Fi hot spots. But most mobile devices now have two radios, one for talking to cell towers and one for talking to Wi-Fi hot spots. In the not-too-distant future, as we move through the city our devices will silently shop around, switching between cellular towers and nearby Wi-Fi hot spots if we linger in one place too long. Wireless carriers in several countries have already deployed such technologies, and Cisco is leading a push for Hotspot 2.0, a new standard for global cellular-to-Wi-Fi roaming. And new smart-radio technologies will increasingly allow our devices to make use of frequencies occupied by older wireless technologies without interfering with existing signals.

Cities concentrate demand for mobile bandwidth, but the tyranny of physics constrains the amount available. They push the data-conveying capabilities of our radio technologies to their limits. Yet while untethered networks are the weakest links in the plumbing of smart cities, they are the most valuable. They free us from the terminals of the industrial age, the typewriters and the telephones that morphed into personal computers but kept us chained to our desks. Instead they allow us to merge with our devices; as sociologist James Katz puts it, they are "machines that become us."[85] This indispensable, intimate, and problematic piece of digital infrastructure will broker our every connection to the systems of the smart city.

We shouldn't be surprised that wireless has won us over. Almost a century ago, at the very dawn of the untethered age, Nicola Tesla saw clearly the world into which we are now moving. A visionary pioneer of electricity and radio technology, Tesla laid the future bare in 1926 in *Collier's* magazine: "When wireless is perfectly applied the whole earth will be converted into a huge brain, which in fact it is, all things being particles of a real and rhythmic whole."[86]

2

Cybernetics Redux

"Representatives and direct Taxes," read the new government's charter in 1787, "shall be apportioned among the several States which may be included within this Union, according to their respective Numbers. . . ." This single sentence of the US Constitution begat the Census, the Census begat IBM, and then IBM begat the modern world. An oversimplification of biblical scale, but allow me to explain.

At less than five thousand words, America's Constitution is one of the world's shortest government charters. But despite its brevity, its authors didn't leave important details to chance. On the very first page, they laid out not only the formula for divvying up seats in the new legislature, the House of Representatives but also the process by which data to feed the calculations should be collected. "Enumeration shall be made within three Years after the first Meeting of the Congress of the United States," the Constitution reads, "and within every subsequent Term of ten Years, in such Manner as they shall by Law direct."[1]

And so, the census was born.

The first count began on Monday, August 2, 1790, just over a year after President George Washington's inauguration.[2] In 1793 the full results were published. In fifty-six pages of elegantly typeset tabulations, *Return of the Whole Number of Persons Within the Several Districts of the United States* described a land of villagers and farmers—barely one in twenty Americans lived in cities and towns in 1790.[3] In New York City, already the nation's largest settlement, dwelled a mere 32,328 persons. This pattern would hold for decades. As late as 1840, just 10.8 percent of the nation's population were city dwellers. But the industrial revolution would change all that. From just 2 million townspeople in 1840, America's urban population grew to over 50 million in 1920, when for the first time they outnumbered the country folk.[4]

As the nation grew, the census grew in scale too. The first census, conducted house by house in 1790, found slightly fewer than 4 million souls in the land. By the tenth count in 1880, some 50 million persons were enumerated.

The scope of data points gathered on each person expanded dramatically as well. Despite the ravages of war, America remained a magnet for immigrants, who showed up in astonishing numbers. From 1850 to 1880, an average of nearly 1.5 million arrived each decade.[5] Alarmed at the unprecedented growth of immigrant ghettos in the cities, a Congress still mostly dominated by rural landowners authorized an expansion of the census's demographic data collection. General Francis Amasa Walker, the economist who had overseen the 1870 census, was tapped again to plan the new survey. He added questions about marital status, birthplace of parents, and length of residence in the United States, and two questions on mental health (Question 18, "Was the person idiotic?" and Question 19, "Was the person insane?" explored apparently obvious contemporary distinctions).[6] More importantly, for the first time the census included a massive survey of the economy, tallying its manufacturing, mining, agriculture, and railroad sectors.[7] The 1870 census reported back in just three volumes; the 1880 report swelled to twenty-two.[8]

The broadened scope of its inquiry overwhelmed the Census Office, then still a temporary group in the Department of the Interior. (The permanent Bureau of the Census would not be established until 1902). Despite tripling the number of staff from the 1870 effort to over 1,500 workers, the full tabulation of the 1880 census lasted seven years.[9] By the time it was completed in 1887, the next count was just three years away. While plans for the 1890 census called for even more staff, many feared that given the accelerating pace of change in the population and economy, "the 1890 figures would be obsolete before they could be completely analyzed."[10] This decoupling of the nation's demographic and economic reality from what could be measured was yet another dimension of the "control revolution" of the late nineteenth century we saw in chapter 1, when "innovations in information-processing and communications technologies lagged behind those of energy and its application to manufacturing and transportation," according to sociologist James Beniger.[11] We were building cities faster than we could count the people pouring into them.

The solution to the young nation's counting problems would come from a former Census Office clerk turned engineer and entrepreneur. Hired to work on the 1880 census, Herman Hollerith hailed from Buffalo, New York.[12] At the Census Office, Hollerith befriended John Shaw Billings, who headed the Division of Vital Statistics. Hollerith and Billings often discussed new approaches to the problem of tabulating the massive piles of data being collected. As Hollerith later recalled, "One Sunday evening at Dr. Billings' tea table, he said to me there ought to be a machine for doing the purely mechanical work of tabulating population and similar statistics. We talked the matter over. . . . He thought of using cards with the description of the individual shown by notches punched in the edge of the card."[13] Punch cards had been used to control machinery since the 1801 invention of the Jacquard loom, a French machine that used thousands of cards to weave extremely complicated patterns in textiles. Their application to data processing held tantalizing potential.

When General Walker left the Census Bureau in 1881 to take over as the president of MIT, he invited Hollerith to join him as an instructor in mechanical engineering. Hollerith soon tired of teaching, however, and found his way back to Washington to work as a patent examiner. Over the course of a busy year, he familiarized himself with the art of patent writing and prior art in punch-card technologies. Supporting himself as a consultant to other aspiring inventors, during the next several years Hollerith began building the tabulating machine first imagined in his conversations with Billings.

Hollerith's machine was remarkably simple to operate. To process a card, which had been punched to record the characteristics of a single individual captured by the Census, the operator simply placed it on a rubber pad, beneath which lay dozens of tiny cups of mercury, and with a handle lowered a swinging array of metal pins. As pins passed through the punched-out holes in the card, they would contact the mercury and close a circuit with a tiny electric motor. On a panel facing the operator, four rows of ten clocklike dials represented the various data items encoded on the card—race, gender, age, and so on. With each pulse, the indicator would advance. Using two rotating hands and a circumference marked off into 100 ticks, it could track sums up to 9,999. From time to time, the operator would read the dials, jot down the totals and reset them to zero.[14]

Compared to manual tabulation, this system was blazingly fast. In June 1890, ninety-six of Hollerith's machines were put to task processing the results of the new census, taken on the second of that month. By the end of the summer, the machines' impact was clear—the raw population count of Washington, DC, was announced on June 28, and New York City a few weeks later on July 18. By the end of August, the full tally of every state, which enumerated over 60 million Americans, was completed. A full statistical breakdown was published in 1892.[15] Hollerith boasted that the bureau could now process a stack of forms the height of the Washington Monument in a single day.[16]

No longer a government employee, and protected by an array of carefully written patents, Hollerith proceeded to fleece his former employer. Instead of selling the machines outright to the Census Office, he leased them. In a particularly usurious move, he developed a rate structure based on the number of cards counted, thus ensuring that the explosion of data gathering would create an equally explosive flow of revenue for his company. With practice, census workers could process five hundred to seven hundred cards per day.[17] At a rate of 65 cents per thousand cards tabulated, that entitled him to over $6,000 a year in fees per tabulator—more than the cost of the machine itself![18] To top things off, he required the government to use his Tabulating Machine Company as the sole supplier of punch cards.

Renting the machines to his customers rather than selling them outright was also part of a calculated effort to protect the machines from copycats. Hollerith's experience as a patent consultant had taught him the urgency of capitalizing on his technological lead.[19] Retaining ownership and handling all of the maintenance and repair of the machines helped Hollerith conceal how the machines worked and the details of their design. But after he jacked up rental rates in advance of the 1900 census, the new Census Bureau made the decision to build its own tabulating machines for the 1910 count.[20] By then, however, Hollerith's success was no longer in question. The market for tabulating machines was expanding rapidly. In the quarter-century following its 1890 debut, Hollerith's invention was put to use in censuses conducted by governments around the world, including Austria, Norway, Canada, and Russia.

But the tabulator's future was in industry, not government. In 1893 Luigi Bodio, the director of Italy's census, prognosticated that "the time will come when the railroads, the great factories, the mercantile houses, and all the branches of commercial and industrial life will be found using the Hollerith machines as a matter of not only economy but necessity."[21] Railroads, the industry at the epicenter of the control revolution, were eager customers. By 1910 Hollerith's subsidiaries were supplying machines to tabulate accounting ledgers and freight manifests throughout North America and Europe.

In 1911, after a lengthy legal and lobbying battle that ended in a termination of his contract with the Census Bureau and probable infringement on his patents by the US government, Hollerith was ready to cash out. At the invitation of Charles R. Flint, the great mergers-and-acquisitions tycoon known as the "Father of Trusts," Hollerith accepted an offer to merge his Tabulating Machine Company with two other firms. With a million-dollar windfall and a cushy salary, he eased into retirement.

The crisis of counting set in motion at the birth of the American republic over a century earlier and brought to a head by the explosive growth of industrial cities was now over. But its solution, Hollerith's tabulating machine, had set the stage for a far greater transformation. For the company formed from that merger, the prosaically named Computing-Tabulating-Recording Company, would pursue an ever-expanding market for information processing throughout the next century. In 1924, under the leadership of Thomas J. Watson, it would take a new name—International Business Machines Corporation.

Big Blue

Fast-forward to 2011, a big year for the company that came to be known as "Big Blue." It's the one-hundredth anniversary of the merger that launched Hollerith's punch-card enterprise on its way to global domination and built a big business processing the big data of government and business. Throughout the twentieth century, IBM's pinstripe-suited engineers personified corporate America. But in 1993, after a long decline driven by growing competition in its mainframe and personal computer businesses, Big Blue hit rock bottom, posting an $8.1 billion operating loss. That year CEO Louis Gerstner Jr., a veteran of RJR Nabisco and American Express, embarked on a radical transformation plan. The new IBM would focus solely on services and integration of large-scale, complex information systems. In 1995 the company abandoned its famously strict employee dress code. A decade later, in 2004, it was ready to jettison the personal-computer division that had so recently defined it.

The new IBM wasn't a staid purveyor of hardware; it was a general contractor for planetary-scale computing. Less than three years before the centennial, in 2008, company chairman Sam Palmisano had launched IBM's Smarter Planet campaign in a speech to the Council on Foreign Relations.[22] If Siemens and Cisco aim to be the electrician and the plumber for smart cities, IBM's ambition is be their choreographer, superintendent, and oracle rolled into one.

While Smarter Planet is a snazzy spin on a new marketing push, IBM has a long history of building truly globe-spanning computer systems. The company boomed after the Second World War, as the consumer economy swelled and carried American firms along with it. Rising international trade, the settlement of the Sun Belt, and increasing leisure time drove a swift expansion in air travel. Much like the management crisis created by the spread of railroads a century before, airlines couldn't keep up with the acceleration of commerce they were enabling. After a chance encounter on a long flight in 1953 between C. R. Smith, the president of American Airlines, and a young IBM salesman, IBM began planning a replacement for the company's archaic paper-based ticketing system.[23] By 1960, drawing directly on its work in the mid-1950s building the massive SAGE (Semi-Automatic Ground Environment) air-defense computer system for the United States Air Force, IBM installed the eponymous SABRE (Semi-Automatic Business Research Environment) for the commercial airline. For the first time, travel agents could call into a specially designed computer center where airline reps could instantly browse available seats. SABRE cut the processing time for reservations from an average of ninety minutes to just a few seconds. As an exhibit at IBM's global headquarters celebrating the centennial proclaimed, "What once took hours could now be done in real-time." A half-century later, after countless upgrades, the fully automated descendant of SABRE still processes over forty thousand bookings per second for dozens of airlines worldwide.

SABRE opened a new chapter in the control revolution. As IBM's corporate historians boasted, "For the first time, computers were connected together through a network that allowed people around the world to enter data, process requests for information and conduct business." It didn't just let American coordinate its operations better;

it revolutionized air travel and set the stage for economic globalization and the urban explosion that revolution would unlock. It presaged "the entire universe of electronic commerce that exploded in the mid-1990s."[24]

The legacy of SABRE is written all over Smarter Planet. Colin Harrison, one of the firm's "Master Inventors," helped launch IBM's efforts to apply its technology to urban problems before retiring in early 2013. As bombastic as IBM's historians may be, Harrison was typical of the company's cadre of inventors, a brilliant yet humble practitioner. His bio lauded many achievements, including leading the development of the first commercially useful MRI system in 1978, but also his "many failed innovations," including an intriguing one named "magnetic bubble memories."[25] For Harrison, the weaving of SABRE-like information systems into everything was an inevitable historical process. "Over the last two decades," he explained in 2011 at a conference in New York, "the planet became wired for transactions. The global supply chains that existed for centuries suddenly became instrumented," fitted out with sensors that could track the movement of people, goods, and money. Manufacturers could track operations and sales worldwide, in real time. Suddenly, suppliers could tap into their customers' mainframes to update delivery schedules. Consumers increasingly got glimpses of this new commercial apparatus, like the package-tracking services provided by carriers such as UPS and FedEx. Streams of sensory data gave companies a holistic new vantage point, according to Harrison—"you could see patterns in what was going on in your particular ecosystem." Frustratingly though, while "during that period this approach to managing was adopted by almost every industrialized domain of human activity," he argued, local governments lacked the networks needed to plug their systems together.[26]

IBM set its sights on government as a huge, untapped market and cities as a particularly high-growth segment. A third control revolution, building on the ones pioneered by Hollerith and SABRE before it, was in the making. But according to John Tolva, IBM's leading evangelist for smart cities at the time, "There was a huge lack of subject matter expertise in cities in the company."[27] To get up to speed, in 2010 IBM tapped into its existing leadership-development program, the Corporate Service Corps, to create the Smarter Cities Challenge. A kind of consulting Peace Corps for smart cities, the program paired teams of pro bono consultants with cities across the world to design solutions that drew on IBM's technology and expertise. A pilot round in 2010 that involved seven cities put IBM engineers on the ground in front of real urban problems. As Tolva explains, it was priceless knowledge. "There was no formal way to get that. It created a couple hundred people," inside IBM, "who know what's going on with cities."[28] Over the next three years the program promised to deliver $50 million in pro bono consulting services to one hundred cities around the world.

By 2011, this missionary strategy was paying off. In early June, lumping in Smarter Cities Challenge and a host of conventional paid engagements with city governments, IBM claimed a knowledge base that spanned over two thousand "Smarter Cities" projects. With these hastily minted credentials in place, the company launched its most ambitious urban solution to date, the "Intelligent Operations Center for Smarter Cities." A kind of mission control center for mayors straight out of NASA, it was the culmination of Harrison's vision of an instrumented approach to city management. Anne Altman, the general manager for IBM's Global Public Sector, made the pitch. The system could "accurately gather, analyze and act on information about city systems and services." It was an all-seeing eye that "recognizes the behavior of the city as a whole." At its heart was a prediction engine offering "deep insights into how each city system will react to a given situation."[29]

Once again, crisis had spurred the creation of a new technology for controlling the city. In April 2010 Rio de Janeiro experienced the worst flooding in its modern history. As a series of sudden, unanticipated rainstorms triggered mudslides, hundreds were killed and tens of thousands made homeless in the hillside slums rising above Rio's tony center. The city's inability to avert the disaster was an embarrassing failure for local officials. Six months earlier, just a few weeks after the city was selected to host the 2016 Summer Olympic Games, the world had watched the televised downing of a police helicopter, caught in the crossfire during a street battle between two rival drug gangs. Rio had suffered a half-century of decline since the government relocated to the newly built capital of Brasilia in 1960. Now, as it prepared to take the world stage, the sprawling city of 6.3 million people seemed more ungovernable than ever.

Mayor Eduardo Paes desperately needed to shake off Rio's lackluster image by taming the city. Soon after the floods, he called in a team of IBM engineers led by Guru Banavar—himself a native of a fast-growing developing world megacity, India's tech hub Bangalore. Paes asked IBM to design a disaster management system that would provide a heads-up view of what was happening in the city and speed the flow of information between different parts of government during a crisis. But he also wanted to prevent disasters in the first place. Could a computer predict the approach of future storms?

IBM already had the answer in Deep Thunder, a high-resolution weather forecasting system that could forecast precipitation up to forty-eight hours in advance. Coincidentally, Deep Thunder had grown out of an earlier collaboration between a team of engineers and meteorologists at IBM and the National Weather Service in 1996, to forecast weather for the Summer Olympic Games in Atlanta.[30] In the intervening years, IBM had continued to improve the software's accuracy. With a resolution of just one square kilometer, the company claimed Deep Thunder was over thirty times more precise than the state of the art at the time. "You can see what's going to happen in the Olympic Village, for example," Banavar boasted at a 2012 Columbia University lecture.[31]

Enamored of the new system, Paes ordered construction of a brand-new building to house it in the neighborhood of Cidade Nova, just a few miles north of Copacabana Beach. The Rio Operations Center is a bunker fit for a president—its *Sala de Controle* (Control Room) houses seventy operators from thirty different city departments. A network of four hundred cameras placed throughout the city pump video to a bank of screens covering an entire wall; a government promotional film brags that it's "the largest screen in Latin America." There's a crisis room linked to the mayor's residence and the national civil defense authorities. The press are cordoned off behind glass in a fishbowl, presumably to be fed a well-spun trickle of news.[32]

What began as a tool to predict rain and manage flood response morphed into a high-precision control panel for the entire city. As Paes boasts in the film, "the operations center allows us to have people looking into every corner of the city, 24 hours a day, 7 days a week." Banavar explains that just a few months into the project, IBM and the city reenvisioned the whole endeavor as more than just a disaster management center. Rather, it would be a way to manage everything in the city—from big happenings like Carnival to everyday events such as concerts. A common operational planning protocol was developed, a preparation checklist that scripts and monitors all of the actions in the days ahead of an event across the gamut of city agencies. From what US Navy strategist Richard Norton described as a city at risk of becoming "feral" in 2003, Paes and IBM have used smart technology to render Rio one of the most meticulously managed on the planet, it seems.[33] "As part of my job . . . I encounter lots of different kinds of cities," says Banavar. "I can't say I've seen any other city that has this level of coordinated governance."[34]

In the spring of 2012, the world got a chance to see just what Rio and IBM had created—a remote-control city. Speaking at the TED (Technology, Entertainment, Design) conference in Long Beach, California, one of the Internet's most visible platforms for big ideas and celebrities, a young, tan, and ebullient Paes played the increasingly prevalent role of nonideological, problem-solving mayor as well as ambassador for a resurgent Brazil's global ambitions. His talk, brazenly titled "The 4 Commandments of Cities," laid out his vision of how to run a city. For the climax, he turned to the screen and dialed up a videoconference with Carlos Roberto Osorio, his point man for urban affairs, back in Rio. For the next minute, Osorio flipped through a dizzying succession of live digital maps and debriefed the mayor on the day's events (it was nearing midnight in Brazil as Paes spoke on the West Coast)—the GPS-tracked movements of the city's garbage truck fleet, current precipitation picked up by the city's brand-new Doppler radar, and Deep Thunder's latest forecast (all clear). To cap off the show, Orsorio served up "a live transmission in downtown Rio for you, Mr. Mayor," beamed from the dash-mounted camera of one of the city's eight thousand buses. "You see, the streets are clear."[35]

Just how effective Rio's Operations Center will be in taming the wild metropolis remains to be seen. Urban security experts with whom I have spoken are skeptical that it will have any significant impact on law enforcement, and technology experts point out that beyond the video streams there has been little investment in new sensor infrastructure to feed real-time data to the center. But as IBM has gotten its hands dirty in real cities, it has learned some valuable lessons about urban politics too. As Colin Harrison explained, after Palmisano's 2008 speech, "Mayors, elected officials, governors . . . people all over the world suddenly wanted to hear more" about IBM's smart city wares. But it soon became clear that looking smart, even more than being smart, was the real force driving mayors into the arms of engineers. "Part of the thinking that you find in elected officials and economic development teams is they want their city to seem modern, to seem Internet-friendly," Harrison continued. "The people they're trying to attract are Internet natives who think of the idea of going to a government office and filling out a paper form as a ridiculous procedure. It needs to be on the Web somehow." Harrison and IBM have absorbed the lesson well. "That was a big surprise to us. We thought that this was going to be about ROI [return-on-investment] models, and the efficiency that we can produce. To some degree it is, but it's economic development and competitiveness that's at the heart of it."[36]

To experienced city watchers, the "look smart" urge is obvious. For decades, enterprising mayors everywhere have lurched from one urban revitalization scheme to the next—sports stadiums, convention centers, and public Wi-Fi—in an attempt to attract talent and businesses. Are the new urban engineers like IBM's Harrison and Banavar stumbling into a hornet's nest of urban policy making, where the variables that need to be optimized are often unclear and routinely fought over with inconclusive results, and where good policies often yield to expedient ones? More importantly, will cities stay committed to these projects, or are control centers like Rio's destined to become tomorrow's white elephants?

Even if mayors stay committed to smart-technology projects over the long haul, will IBM's newfound love for cities last? Fifty years ago, IBM's leadership decamped from its headquarters in midtown Manhattan to a wooded ridge in Armonk, New York, in 1964, taking thousands of jobs and a big chunk of New York City's pride with it. The Googleplex of its day, the Armonk campus was a calculated withdrawal from the growing problems of America's most important city. In the years since, IBM has amassed a tremendous arsenal of talent and technology to tackle urban problems. And while the actual engineers building IBM's technologies are on front lines all over the world, it is worth pondering whether cities should blindly follow a company that takes its best minds and hides them away in a posh suburb.

Mirror Worlds

The city control room IBM built in Rio shouldn't surprise us. In 1991 Yale University computer science professor David Gelernter foretold all of it in stunning detail. "This book describes an event that will happen someday soon: You will look into a computer screen and see reality," begins his book *Mirror Worlds*. "Some part of your world—the town you live in, the company you work for, your school system, the city hospital—will hang there in a sharp color image, abstract but recognizable, moving subtly in a thousand places . . . fed by a steady rush of new data pouring in through cables . . . infiltrated by your own software creatures, doing your business."[37] It was a vision so all-encompassing and transformational that it spurred mail bomber

Ted Kaczynski to break a six-year hiatus in 1993, and dispatch the incendiary missive that narrowly missed taking Gelernter's life.

Mirror Worlds foretold with astonishing accuracy the way sensing, networking, computation, and visualization are converging in our world today. But what's really interesting is how over and over Gelernter used cities to illustrate the power of tools that capture vast complexity in real time. It starts on the first page of chapter 1: "Suppose you are sitting in a room somewhere in a city, and you catch yourself wondering—what's going on out there? What's happening? . . . At this very instant, traffic on every street is moving or blocked, your local government is making brilliant decisions, public money is flowing out at a certain rate, the police are deployed in some pattern. . . . This list could fill the rest of the book."

Gelernter's vision grows even larger when you add the dimension of time. Imagine your local Chinese takeout joint and all the orders flowing in, the tens of millions of rice pails delivered since it opened decades ago, all of the accumulated history of mundane transactions that happened there. Or, in some old-timers' corner tavern, all of the glasses raised in a century's libations. Cities are deeply complex, built up through a vast array of small activities that accumulate over time. What if we could record, preserve, analyze, and visualize that detail?

Mirror Worlds described how those images would come to life, not in the form of a machine intelligence that could make sense of it all, but a new kind of all-seeing eye that would give humans the ability to do so. "Mirror worlds," Gelernter wrote, are "scientific viewing tools" that focus "not on the hugely large or small, but on the *human-scale* social world of organizations, institutions and machines; promising that same vast *microscopic, telescopic* increase in depth, sharpness and clarity of vision." As powerful as zooming into detail was, however, for Gelernter it was a red herring. The real power of mirror worlds wasn't from insight. What he was after was "topsight . . . what comes from a far-overhead vantage point, from a bird's eye view that reveals the *whole*—the big picture; how the parts fit together."[38]

As interesting as his descriptions of mirror worlds are, Gelernter's critique of them is even more fascinating. At the very outset of the book, he declares "the *social* implications of these software gizmos make them far too important to be left in the hands of the computer sciencearchy."[39] Yet not until the book's bizarre epilogue do we hear another critical word, and it comes in the form of a schizophrenic fictional conversation between Gelernter's alter egos, a musician named Ed and an electrical engineer named John. On the pages that follow, Ed and John give voice to Gelernter's alternating excitement and misgivings about a future dominated by mirror worlds. Perhaps he wanted to distance himself from the downsides of mirror worlds he felt obligated to disclose. Perhaps he thought they would be taken more seriously if he did so.

Gelernter quickly gets to the point: humanity will become dependent on mirror worlds, and that will destabilize society. Ed, the critic, makes the case by explaining how the invention of the stirrup spurred an arms race in Europe, bringing about the professionalization of mounted warfare, and the feudal system needed to finance it. Similarly, mirror worlds would spur an informational arms race. Whoever could assemble a mirror world would trounce those who could not. The result would be upheaval. Mirror worlds were "a centrifuge . . . designed to stratify society based strictly on a person's fondness for playing games with machines."[40]

But it wasn't just the material basis of society that was at stake in mirror worlds; it was our very minds, our individual and collective process of reasoning. Speaking through Ed, Gelernter writes, "It's not that I *distrust* the software guys who design and build them. . . . They'll take good care of us. And that's *just* the problem. Serfdom means, above all, not slavery—slavery is slavery; serfdom is merely utter *dependency*—I don't understand these things but I rely on them, not just for *convenience* but in order to carry out my *thinking*!"[41]

As they rush to build their own mirror worlds, what are cities like Rio de Janeiro giving up? As we have seen, the mere appearance of control, the appearance of doing something about the city's problems with technology, is becoming key to economic survival in a world where cities compete for talent and investment. Yet even as Gelernter frets over dependency on sensor-powered simulations, IBM's Colin Harrison sees it as simply another risk to be managed. "Society rides a number of tigers," he explained to me, "where we've introduced a technology and we can't take it away again. Chemical fertilizer was certainly one of those. Electricity is another one."[42] Harrison sees the deployment of smart-city information systems as yet another irreversible layering of technologies atop these earlier inventions. And IBM holds the high ground—swapping out one company's mirror world for another's isn't even an option. Does that make Rio not just a slave to its mirror world, as Gelernter feared, but also to the company that designed and operates it?

While mirror worlds like Rio's are today designed only for managing cities, the topsight they deliver will be utterly seductive to anyone charged with planning them. But history suggests that these kinds of technologies can be dangerous. As urban planning scholar Tom Campanella explains in *Cities From the Sky*, the invention and widespread use of aerial photography has inflicted untold damage on cities. First, it was used to systematically survey cities for the purpose of planning and targeting bombardments during World War II. Afterward, it provided the perch from which modernizing mayors, developers, and urban planners reigned as virtual gods. Detached from the life of the street, this new perspective inspired soulless designs for modern, mega-scale cities.[43]

Mirror worlds may also create opportunities to improve city planning, by improving our understanding of how cities change over time. Aerial photography showed us only the muscular and skeletal structure of the city. Examining smart cities' sensors will reveal their circulatory and nervous systems. For the first time we'll see cities as a whole the way biologists see an organism—instantaneously and in excruciating detail, but also alive. Today we see them the way astronomers see a heavenly body—as it was, some time ago, light-years in the past. Because of this lag, we plan the future for cities that have already changed into something else.

Still, better topsight won't tell us much about lives of those who actually live in the city. By editing out the "chaotic multi-sensual reality . . . the sights, the sounds, the smells, the character of the people," as Gelernter described it, mirror worlds leave out the subjective reality of city dwellers themselves. What can better topsight tell us about the street-level insights of everyday people? It might just distract us from those voices.

One hot summer morning in 2011, twenty years after *Mirror Worlds* hit the streets, I lay reading it under a tree across the river from Manhattan. Toward the end of his alter egos' debate, Gelernter's thought experiment finally reaches its conclusion—mirror worlds would end the philosophical struggle between the rational objectivism of science and the irrational emotionalism of romanticism that stretched back to the Enlightenment in the eighteenth century. The romantic's worldview, driven by nature and human sensuality, "is dying, because it's inefficient. It doesn't *produce* anything. Except maybe a vague sense of well-being; but so does a bottle of wine." As I had learned about cities over the years, I would often daydream mirror worlds of my own, trying to imagine all of those rich happenings in the metropolis around me. I put the book down and indulged myself to build one last mirror world of my own, trying to see in my mind's eye the cars moving along the West Side Highway, the fares of all the taxis in Manhattan adding up, the bits flowing in on cables under the river. Someday soon, IBM will switch on a real mirror world of Manhattan and destroy the wonderful ephemerality of it all for me, forever. As for Gelernter, "The future is clear. Know everything, feel nothing." Romanticism was on life support. "And Mirror Worlds have the stuff to kill it."[44]

The Psychohistorians

Gelernter foretold the mirror worlds IBM is installing in cities around the world. But the first attempts to use computers to simulate, manage, and plan cities date back to the Cold War. In 1951, Isaac Asimov, the legendary science fiction writer, opened his sci-fi classic *Foundation* with a scene that is familiar to anyone who keeps a tablet computer at hand: "Seldon removed his calculator pad from the pouch at his belt. Men said he kept one beneath his pillow for use in moments of wakefulness. Its gray, glossy finish was slightly worn by use. Seldon's nimble fingers, spotted now with age, played along the files and rows of buttons that filled its surface. Red symbols glowed out from the upper tier."[45]

In the novel Seldon leads a renegade sect, the "psychohistorians," who have developed a "branch of mathematics that deals with the reactions of human conglomerates to fixed social and economic stimuli."[46] Wielding advanced statistics, psychohistorians aspired to predict the future. And Asimov had a knack for inspiring readers to make his visions of the future come true. *Foundation* urged an entire generation to try to tame society with math and computers. Paul Krugman, winner of the Nobel Prize in economics, once said "I wanted to be a psychohistorian when I grew up, and economics was as close as I could get."[47]

Much like economics today, Asimov's psychohistory was a dismal science, riddled with guesswork. In the opening pages of *Foundation*, Seldon indoctrinates Gaal Dorneck, a new apprentice, in the art of psychohistory:

> He said, "That represents the condition of the Empire at present."
>
> He waited.
>
> Gaal said finally, "Surely that is not a complete representation."
>
> "No, not complete," said Seldon. "I am glad you do not accept my word blindly. However, this is an approximation which will serve to demonstrate the proposition. Will you accept that?"[48]

Asimov's depiction of psychohistory was inspired by the new field of cybernetics. Along with nuclear fission and rocketry, the costarring technologies in the science fiction of the day, automated control systems were one of the great technological leaps of World War II. Led by Norbert Wiener at MIT, cybernetics built on wartime research in antiaircraft targeting techniques that used past observations of flight trajectories to improve predictions of an aircraft's future position. Cybernetics took the idea of using sensing and feedback to optimize performance and extended it to the universe generally. To cyberneticians, everything—machines, organizations, cities, even the human mind—could be seen as a system, a balanced network of things connected by information flows. The components of any system, and the flows between them, could be represented as a set of equations that together could replicate the behavior of the whole, they believed. With this mathematical "model," an analyst could make predictions simply by changing the inputs and observing the ripple impacts propagate throughout the simulation. It was an immensely powerful idea. Cybernetic thinking inspired new directions in engineering, biology, neuroscience, organizational studies, and sociology.

Cybernetics underpinned the plotline for *Foundation*, but advances in computing provided the props. Just weeks before the 1945 American nuclear strikes on Hiroshima and Nagasaki, Vannevar Bush published a seminal article in *The Atlantic* that laid out a road map for the computer age. Bush was a technological authority without equal, an MIT man who during World War II had directed the entire US scientific effort, including the Manhattan Project that developed the nuclear weapons used against Japan. Like Asimov's psychohistorians, who wielded tablet computers as cognitive prosthetics as they built their socioeconomic simulations, Bush believed that the new thinking machines would liberate the creative work of cyberneticians from the drudgery of computation. "The advanced arithmetical machines of the future will be electrical in nature," Bush predicted, "and they will perform at 100 times present speeds, or more." A mathematician, he wrote, "is primarily an individual who is skilled in the use of symbolic logic on a high plane. . . . All else he should be able to turn over to his mechanism, just as confidently as he turns over the propelling of his car to the intricate mechanism under the hood." The essay is often cited for its description of a hypothetical device Bush called the "memex," a startlingly prescient depiction of the Web browser. But Bush also foresaw the application of computers to understanding entire societies. "There will always be plenty of things to compute," he wrote, "in the detailed affairs of millions of people doing complicated things."[49]

Cybernetics provided a theoretical wrapper for the more mundane field of operations research, which also grew out of wartime planning and applied the new science of systems to the simulation and planning of large organizations. These ideas were deeply embedded in the design of massive, networked organizations like the air defense system coordinated by SAGE. But it didn't take long for cyberneticians to turn these techniques, and the new power of computers, to the problem of America's cities. Like Seldon, they made hasty approximations as they rushed to twist a complex urban reality into a computable set of equations. But unlike the psychohistorians in *Foundation*, whose doomsday prophecies were fulfilled by the story's end, the real world cyberneticians never succeeded in building a machine that could predict the city. In fact, they failed. And that failure had terrible consequences.

As a grad student at MIT in the late 1990s, one wintry afternoon in the library I stumbled upon a curious book called *Urban Dynamics*, by Jay Forrester. I was spellbound to discover in its musty pages an entire science of cities, seemingly forgotten for decades, laid out in objective prose and logical cybernetic flowcharts. Like Wiener, Forrester was a professor at MIT and had also worked on targeting systems during the war. But his subsequent interest in cybernetics was more practical. During the 1950s, Forrester co-led the design of SAGE, a masterstroke of cybernetics that linked up dozens of control bunkers with over one hundred radar stations throughout North America.

Forrester's experience building SAGE taught him that engineering wasn't the biggest obstacle to building big, complex technical systems. The real challenge lay in managing the people and organizations who would use them. Humans, it turned out, were far harder to understand and control than machines.[50] Beginning in 1956 at MIT's new Sloan School of Management, he quickly became one of the leading lights in operations research. While cyberneticians like Wiener debated the nature of the universe elsewhere on campus, Forrester was more interested in actually designing really complex things. He developed techniques for mathematically modeling industrial systems, focusing on how feedback loops and time delays governed flows and stockpiles of resources and products. The culmination of that work, *Industrial Dynamics*, was published in 1961. It analyzed the workings of a General Electric plant in Kentucky, laying the foundations for modern supply-chain management.[51]

Having mastered the corporation, Forrester looked for other complex systems to which he could apply the cybernetic tool kit he now called generally "system dynamics." When former Boston mayor John Collins was appointed as a visiting professor of urban affairs at MIT and, by sheer coincidence, moved into the office next door, Forrester seized the opportunity.

Forrester wasn't the first to get the idea that computer models could be used to understand cities. The success of systems engineering in the massively complex defense and aerospace sectors held out the hope that it was up to the task of city management. It was a time of great anxiety about the future of American cities. Summertime riots had become an almost annual event in inner cities, as jobs and the well-to-do fled for new suburbs. As Forrester wrote in the introduction to *Urban Dynamics*, published in 1969, "The plight of our older cities is today the social problem of greatest domestic visibility and public concern."[52]

Using Collins's connections, he canvassed experts on a range of urban issues. He developed equations that described how various parts of the city operated—housing and labor markets, for instance—and how they interacted with each other. These relationships were programmed into a computer to create a simulation that purported to explain how cities grow, stagnate, decline, and recover.[53]

Rather than studying a particular city, *Urban Dynamics* was an attempt to abstract a generic system model of cities. But the book confounded urban policy makers, not just because of its lack of grounding in an actual place but because of its counterintuitive conclusions. Forrester's generic city started in a "stagnant condition" that seemed to characterize most big US cities at the time—a stable equilibrium of high unemployment, a surplus of slum housing, and a shortage of housing for professionals. But setting the model to simulate prevailing urban policy, such as job training for the unemployed and direct federal aid to cities, actually resulted in worse outcomes. Even more surprisingly, the results argued in favor of the policy of demolishing slums and replacing them with high-end commercial and residential buildings, a tactic that by the end of the 1960s was already highly controversial. Nevertheless, Forrester had an "unflinching confidence" in his methods and the results, as a book reviewer in the *Journal of the American Institute of Planners* put it.[54] He casually excused the book's lack of reference to any contemporary work in urban studies. "There are indeed relevant studies on urban behavior and urban dynamics," he wrote, "but to identify these is a large and separate task." With no formal training in urban planning, based solely on his computer simulation, Forrester recommended the demolition not only of slums but of federally subsidized public housing as well, which the model showed became poverty traps for their inhabitants. While the ghettoization of the poor in housing projects is now widely recognized, it was obvious failures like the disastrous Pruitt-Igoe complex in Saint Louis (which was torn down in the 1970s) and painstaking fieldwork by a generation of social scientists that made the case in the end.[55]

Urban Dynamics was perhaps the most ambitious effort of that generation of computer-based urban simulations. But it came at the tail end of a decade of failures to apply systems analysis to urban problems. As historian and sociologist Jennifer Light explains in

From Warfare to Welfare: Defense Intellectuals and Urban Problems in Cold War America, much as IBM turned to cities for new business during the 2008 financial crisis, the defense industry began looking for new markets for military computer technologies almost as soon as they were invented. As early as 1957, connections were being drawn between the similarities of military planning and urban planning.[56] Uncertain how long the Cold War would sustain defense spending, Light argues, the think tanks "decided that the survivability of their organizations depended on finding ways to transfer their innovations beyond military markets." In the late 1950s, defense contractors such as TRW and RAND began publishing studies in urban and public administration journals, Light recounts, "suggesting how techniques and technologies from military operations research such as systems analysis and computer simulations might offer a new direction for city management."[57]

The results were less than impressive. In the early 1960s, as part of its federally funded Community Renewal Program, the city of Pittsburgh attempted to develop computer simulations that would forecast the impacts of public spending decisions about transportation, land use, and social services. Almost immediately, problems appeared. One program that sought to measure the impact of housing clearance for an expressway produced nonsensical results.[58] Rather than expand the city's capacity and inform better decisions, technology constrained thinking. As Light explains, Pittsburgh's planners "realized they were shaping their questions and problems to fit what could be modeled . . . yet rather than characterize this as a flaw of simulation techniques, they used this finding to justify why one would want to use them." Captured by the computers' limits, they argued that simpler models were better. In their words, complex models that were "photographic reproductions of reality . . . would be so complicated that they would be of little, if any use."[59] With nothing usable to show for its modeling efforts, in 1964 the city fired the project's director and declined to apply for an extension of its federal funding for the effort.[60]

Like the psychohistorians, the urban modelers of the 1960s had a maddening habit of relying on approximations, a practice that had devastating consequences in New York. As Joe Flood describes in his 2010 book *The Fires*, facing a rising wave of blazes and union demands for more resources, in 1969 New York City fire chief John O'Hagan turned to the New York City-RAND Institute, a partnership with the think tank that Mayor John Lindsay had formed little more than a year earlier. It was a bold attempt to apply cybernetic thinking to the operations of local government—as Lindsay described it, the "introduction into city agencies of the kind of streamlined, modern management" that defense secretary "Robert McNamara applied in the Pentagon with such success in the past seven years." Focusing on just a single measure of fire company performance, response time, RAND developed a computer model of the city's firefighting system.[61] Despite the RAND analysts' own misgivings about the usefulness of response time, it was the easiest indicator to measure reliably, and was less variable and therefore simpler to model. As Flood explains, "RAND made a fateful choice: gather the response-time data, model it to the best of their abilities, and put their concerns about response time's shortcomings to the side."[62]

The assumptions, and the distortions they created, compounded from there. RAND's model also assumed that fire companies were always available to respond from their firehouse, which in actuality was "a rarity in places like the Bronx, where every company in a neighborhood, sometimes in the entire borough, could be out fighting fires at the same time," Flood explains. Another half-witted shortcut left out the paralyzing impact of gridlock; "in the most congested city, traffic played no role in response times, rigs able to cruise through Midtown Manhattan at rush hour at the same speed as through Queens at midnight."[63] Politics distorted the model too, without much pushback from its designers. As RAND's Rae Archibald told Flood, "If the models came back saying one thing" and fire commissioner John O'Hagan "didn't like it, he would make you run it again and check, run it again and check."[64] During a wave of budget cuts in

1971, RAND's model counterintuitively recommended shuttering several of the busiest fire companies in the city, based solely on its calculations of response times.[65] The resulting closures were concentrated in poor areas of the city; the demands on remaining fire companies soared, and the Bronx (and several other neighborhoods) burned. Flood puts the tally of persons displaced by the fires at more than a half-million.[66]

By the mid-1970s, in every domain of urban planning and management to which computer modeling had been applied—generic system models like Forester's, land use and transportation models like Pittsburgh's, and even relatively narrowly focused operational models like the one built by RAND for the New York City Fire Department—serious doubts about its effectiveness had been raised. By the mid-1970s, planning scholars moved swiftly away from their earlier embrace of such all-encompassing, predictive city simulators. In 1973 Douglass Lee's "Requiem for Large-Scale Urban Models" sounded their death knell in the pages of the *Journal of the American Institute of Planners*. Then a professor of urban planning at the University of California, Berkeley (today he still works on models for the US Department of Transportation's Volpe National Transportation Systems Center), Lee had studied the Pittsburgh model up close while working there.[67] The article was a scathing indictment, calling out "seven sins" of large-scale models—hypercomprehensiveness, grossness, hungriness, wrongheadedness, complicatedness, mechanicalness, and expensiveness. But Lee reserved his most searing commentary for Forrester, the MIT professor who "buries what is a simplistic conception of the housing market in a somewhat obtuse model . . . then claims that the problem cannot be understood without the irrelevant complexity."[68] While the Pittsburgh modelers had dumbed their model down to make it tractable, Forrester had embellished his to make it look more sophisticated.

City planners relegated cybernetics and system dynamics to the doghouse for the better part of thirty years. The Urban Systems Laboratory at MIT closed doors in 1974 for lack of funding. Louis

Edward Alfeld, who directed Forrester's urban research in the early 1970s, wrote in 1995, "The past twenty-five years have not treated urban dynamics kindly. . . . It has become a curiosity, a relic of the past that few have heard of and most dismiss."[69] The same year, in a retrospective on "Requiem," Lee noted that "modeling is mostly a cottage industry, not much different from what it was ten or twenty years ago. Despite upheavals in planning and the massive changes in computing technology, the role of [large-scale urban models] remains unresolved. That [they] are alive and well may be fine for the modelers, but is it of consequence to anyone else?"[70]

System modelers were cast out from the city in the early 1970's, but the discipline blossomed in the private sector, where it was effective at tackling the analysis of less complex systems than an entire city.[71] As it turned out, their exile would not be permanent.

The year 2011 witnessed cybernetics redux when IBM resurrected urban dynamics and installed it in Portland, Oregon, a city of some half-million people. While the simulation efforts of the 1960s had to cope with severely limited computers and data-collection capabilities, with virtually limitless processing power and vast stores of digitized data at its disposal, IBM developed a computer model of Portland that dwarfed Forrester's. "System Dynamics for Smarter Cities," as the apparatus was blithely named, wove together more than three thousand equations. Forrester's had used just 118 (only 42 of which, a subsequent analysis determined, really shaped the results).[72] On a website used to interact with the model, diagrams reminiscent of those in *Urban Dynamics* dissected the city into a spaghetti-like tangle of interacting variables. It was as if someone sauntered into an IBM lab, dropped off a copy of the moldering book, and said, "Give me one of these." And in so doing ignored forty years of painstaking learning and progress in urban modeling and simulation.

Where IBM's Deep Thunder simulation in Rio predicted rainfall up to forty-eight hours in advance, the one it built for Portland, grinding on ten years of historical data, was meant to project years into the future and inform long-term planning (much as Pittsburgh's model in the early 1960s was meant to inform a master plan for 1980).[73] Planners could ask the program questions by toggling different controls. "How would transportation policy investments affect K-12 education? How would parks and land-use decisions effect greenhouse gases?" explained Joe Zehnder, Portland's chief city planner at the time.[74] The software would spit out predictions in response. IBM touted it as a "decision support system," a tool to help policy makers explore the ripple effects of different options, and the interdependencies of different systems in the city.[75]

The idea to resurrect urban dynamics came from Justin Cook, an IBM strategist who himself was a graduate of the Sloan School where Forrester had once taught. By 2009 IBM had accumulated a deep reservoir of systems modeling knowledge from its work with industry. Cook saw an opportunity to apply it to the company's new Smarter Cities initiative. Looking for a pilot, he said, "I decided that Portland might make a very good candidate . . . they were at the very beginning stages of working out a twenty-five-year plan." Late in 2009, he approached Mayor Sam Adams, a leading advocate of sustainable urbanism, with a proposal for what was not to be a traditional consulting engagement, but rather what Cook described as a "joint research project."[76]

Although "there was a good deal of skepticism" among local economists and planners "that this could be done because of the inherent complexity of a model like that," according to Zehnder, the project moved forward anyway. Over the next year, IBM worked with Zehnder's office and the local experts to develop the map of equations and an arsenal of historical data that would power the simulation.[77] With the help of San Francisco-based Forio, a developer of business simulations, IBM began to weave a spiderweb of relationships that quickly ballooned to over seven thousand equations (a number that was deemed too complex), was pruned back to six hundred (too simple), and then eventually built back up to the roughly three thousand contained in the final revision.[78]

Given the suspect track record of system dynamics in cities, IBM's decision to bring cybernetics back to urban planning was less reckless than it at first appears. As Zehnder described how the model was constructed, with a series of workshops and iterative designs, it appeared to be a vast improvement over Forrester's process—which seems to have taken place mostly behind the closed doors of his laboratory after a cursory round of interviews with ex-mayor John Collins's buddies. Although it's not clear whether Cook was aware of the criticisms of *Urban Dynamics* before the project began, local experts raised those old concerns immediately. But, as Cook told me, the context for building and using systems models of cities had changed dramatically: "Now you can actually take a model like this and put a web interface on it and let people interact directly with the tool and even change some of the assumptions that are in it. That was pretty powerful." In defense of system dynamics itself, the method "is very explicit about the relationships," he convincingly argues, "instead of being a black box where people can't see the logic. We thought this was especially important for working with cities and the constituencies that they could see into the guts of this and make sense of it."[79]

In the end, however, the Portland model, like Forrester's, had little impact on policy. Unlike Forrester's model, which spat out absurd contradictions that actually did stimulate debate, IBM's predictions in Portland were reliably dull. Its greatest revelation, much ballyhooed in the company's public relations campaign for the project, was a strong correlation between the adoption of pro-bicycle municipal policies and a decline in obesity. But no one in bike-obsessed Portland needed three thousand equations to know that. When I asked Zehnder what role, if any, the model played in the planning process, his response indicated that it was largely a sideshow. "It proved . . . to be something where we weren't really going to be able to maintain or use it—in a way that people were going to have confidence in—to illustrate these relationships."[80] But as Cook explained, and Zehnder concurs, the real benefit of building the model was teaching people that cities are "systems of systems," to use a phrase Colin Harrison has advanced to explain IBM's approach to the complexity of smart cities. As Zehnder explained, the result was "an increased awareness that, like all cities, [Portland] operates in silos," a bureaucratic term for government departments that don't cooperate effectively. "Their decisions affect other parts of the city."[81]

After Portland, Cook turned the software over to another business unit within IBM to market it to other cities. At the time we spoke in late 2012, there were still no takers. The challenge for models like these in the future will invariably lie in better balancing the value gained (which is still too small) with the effort required by the city to maintain and operate it (still too high).

While IBM's efforts in Portland may have avoided the kind of devastating consequences that resulted from the first wave of systems models of cities in the 1960s—partly because IBM built the model in a more responsible manner, and partly because the planners chose to ignore it—the project has again raised important, lingering questions about the value of computer simulations of cities.

Michael Batty founded and directs the Centre for Advanced Spatial Analysis at University College London, one of the world's leading centers for urban modeling. Over a career that began at the University of Manchester in 1966, Batty has advanced the science of simulating cities through its dark ages, connecting those ambitious, failed early efforts with today's more modestly successful ones. In a 2011 article, "Building a Science of Cities," he explains the limits of systems models, and why they were abandoned in the first place. Systems models like Forrester's, Batty argues, "treated cities as organised from the top down, distinct from their wider environment which was assumed largely benign, with their functioning dependent on restoring their equilibrium through various negative feedbacks of which planning was central." Forrester's methods for analyzing systems assumed a closed loop—everything that mattered to the system's behavior was contained in the equations. There was no external environment, or at least not one that mattered. And it largely saw the process of change as a shift from one steady state, or equilibrium, to another, in response to some directed action. "As soon as this model was articulated, it was found wanting," Batty counters. "Cities do not exist in benign environments and cannot be easily closed from the wider world, they do not automatically return to equilibrium for they are forever changing, indeed they are far-from-equilibrium. Nor are they centrally ordered but evolve mainly from the bottom up as the products of millions of individual and group decisions with only occasional top down centralised action." In the decades since Forrester, the science of complex systems had taken a 180-degree turn. Mechanical metaphors had been replaced by biological ones, grand design by evolutionary processes, closed loops by open fields of influence. Hammering home the point, Batty concludes, "What has been realised in the last 50 years, is that this notion of systems freely adjusting to changed conditions is no longer valid, in fact it never was."[82] Ecologists had long ago discarded the notion of stability in living systems. But this central tenet of cybernetics, equilibrium, remains firmly embedded in the popular imagination of how human and natural systems behave.[83]

We can only hope that IBM and other would-be urban system modelers will learn from the missteps of cybernetics redux in Portland. Despite theoretical flaws and practical failures, Forrester and his disciples never gave up hope that their methods would one day revolutionize social science and policy analysis. More than twenty years after the publication of *Urban Dynamics* was met with harsh criticism, an unrepentant Forrester proclaimed the universality of systems; he lamented in a 1991 speech that "There is an unwillingness to accept the idea that families, corporations, and governments belong to the same general class of dynamic structures as do chemical refineries and autopilots for aircraft."[84] If there were shortcomings to systems models of cities, his disciple Louis Alfeld argued in 1995 they were "limited detail and limited resources . . . [which] can be overcome by new hardware and software technology."[85]

In the meantime, a range of modeling techniques has supplanted system dynamics in urban research, including many Batty has helped develop. They are showing promise where system dynamics failed. Where system models tried to replicate macrolevel behavior, new techniques such as agent-based simulation use fast parallel-processing computers to simulate the minute interactions between individuals (or "agents") at the microlevel iteratively over time, and calculate the aggregate impact of millions of simultaneous actions. One of the largest such models, developed at ETH Zurich, one of Europe's leading technical universities, in 2004 successfully replicated the actual traffic patterns of Switzerland's 7.2 million inhabitants. And unlike Forrester's static equations, each individual agent can learn and adapt to changing conditions, such as congestion, from one cycle to the next just like real people.[86]

Decades of research lie ahead before we can hope to create software simulacra of cities that approach the psychohistorians' standards of society-scale prediction. And on top of the challenges that have dogged past efforts, new challenges for urban models are on the horizon. For starters, the very same apparatuses that will feed big data into future models—mobile phones, instrumented infrastructure, and digital transaction records—are changing the way cities actually function. As Batty explained to me: "That's the other side of the coin. New communications systems at the local level are actually changing how we communicate. It's not just a question of measuring things that we always did. It's a question of new things emerging. There is a lot of new interaction going on . . . building dynamics into the city that we've never had a hold on at all."[87] Even if IBM's model is perfect today, tomorrow it could be out of date, as new technologies allow us to rewire behavior at the individual level. Even if we can measure the movement of every person in real time, all we'd have is topsight, the big picture. Without an understanding of why individuals are, say, changing the time of day they commute (based on real-time traffic reports beamed to their phone perhaps), we can't accurately simulate their behavior. The models break. It's even possible that those new behaviors are evolving so fast that even our revised assumptions will be out of date by the time they're programmed into the simulator. Theory will lag reality, and the way cities work might actually get weirder and more complicated far faster than we can decode and model it.

Then there's the risk that by measuring something, we change it—a kind of observer effect for social science. Typically understood, the observer effect describes how instruments used to quantify dimensions of the physical world can actually alter the conditions they seek to size up.[88] In experimental physics, it means that to measure the velocity of one subatomic particle, you've got to bounce another one off it like a billiard ball, thus changing the thing you're trying to measure. In electronics, a voltmeter actually becomes part of the circuit itself.

This principle is so fundamental to scientific measurement that Asimov even incorporated it as a central axiom of psychohistory. "The human conglomerate," he wrote in *Foundation*, must "be itself unaware of psychohistoric analysis in order that its reactions be truly random."[89] Will people in a sensed and modeled city behave differently, either by choice or because some plan or policy based on the model directs them to? Either way, it could break the model's assumption and reduce its results to nonsense.

If we assume for a moment that all of these obstacles can be overcome, we are still left to ponder whether better computer models will lead to better cities. The technocratic, top-down style of city planning that gave rise to earlier models is today as archaic as their computer code. Citizens now expect to see, participate in, and even initiate plans. But complex computer models will bring back technocratic opacity, "black boxes" where, as Douglass Lee put it, "What goes in and what comes out are known exactly, but the process by which one is transformed into the other is a mystery."[90]

A far bigger risk is that public officials will accept the advice of these black boxes unquestioningly. As Colin Harrison recounted, early in the Portland model's development, the mayor "formed an idea in his mind of what this model was going to be able to do . . . the planners thought that he was viewing this model as a kind of oracle. He could ask any planning question of the oracle, and it would tell him what the right thing to do was. The planners got very, very nervous about this, and we had to work through this to make sure that he understood that models aren't oracles."[91] It was a surprisingly responsible response by IBM.

Gelernter saw this as perhaps the greatest risk of mirror worlds—that we would mistake them not as reflections or representations, but as reality. Toward the end of the *Mirror Worlds'* epilogue, his alter ego Ed rants: "I can *in fact believe* that a Mirror World would suck life from the *thing it's modeling* into itself, like a roaring fire sucking up oxygen. The external reality becomes just a little bit . . . not superfluous; second-hand. . . . Couldn't it happen that, instead of the Mirror World tracking the real world, a subtle shift takes place and the real world starts tracking the Mirror World instead?"[92]

Computer simulations seduce precisely because they replace the complexity of the real world. The video game SimCity is addictive because of the simplicity of its underlying model—players quickly figure out how to win by exploiting its predictable dynamics (in fact, the design of early versions was directly borrowed from *Urban Dynamics*. Following trends in research, SimCity 2013's GlassBox simulation engine now uses a sophisticated agent-based model).[93] But even the best mathematical models of real-world phenomena are always approximations. Newton's laws made sense for centuries until physicists began looking at the very small scale of matter inside the atom. There a weird new physics reigned and a new model, quantum mechanics, had to be developed to give a better (but still not perfect) approximation of reality.

When I first learned of IBM's work to bring back urban dynamics in Portland, I set out to unmask a villain. What I found was a company perhaps ignorant of a long-buried past, yet willing to listen to experts and learn from its missteps. IBM now knows the political limits of system models of cities. But I wonder if the company has absorbed the more fundamental lesson on their practical limits.

Cybernetics redux in Portland was premised on the notion that bigger data, bigger computers, and bigger models were the remedy to Forrester's shortcomings. It's a familiar, but hollow refrain. As Lee wrote in 1973, "Despite the many-fold increases in computer speed and storage capacity . . ." in the 1960s, "there are some researchers who are convinced that it has been the hardware limitations that have obstructed progress and that advances in modeling are now possible because of larger computer capacity. There is no basis for this belief; bigger computers simply permit bigger mistakes."[94]

A Tale of Two Models

IBM's Banavar is sanguine about the centralization of power in Rio's Intelligent Operations Center. "For better or worse," he reflects, "we have given a lot of power to our municipal governments." There is clearly a case to be made that the urgency of urban problems, especially those faced by mayors in the developing world, justifies arming them with powerful new software and richly detailed information. "I strongly believe we should give them the right tools and the right data to be better managers," Banavar says.[95]

But if we share Gelernter's concerns, we should worry that the mirror world Rio's mayor Eduardo Paes has created in cahoots with IBM will tip the balance of power decidedly in his favor. For now, Paes claims to act in the people's interest. "Every day since I joined the city government," he expounded in the promotional film produced for the Rio Operations Center, "I have dreamed of having this space for the people . . . for people to know that they are being cared for." Paes doesn't hide his paternalistic philosophy of governance; neither is it completely out of place in Brazil. But as IBM exports this new technology and management playbook to the rest of the world, can the ideology from which it was spawned be left behind? And what happens when progressives leave power, and the new tool is turned by autocrats against the people instead?

"Brazil is not for beginners," the country's most famous songwriter,

Tom Jobim, the bossa nova genius who gave the world "The Girl From Ipanema," once said. One wonders at the wisdom of picking such a complicated place to launch a high-profile showcase for IBM's smart city ambitions. The fashioning of Brazil's cities has been a story of chaos, dissent, and grassroots improvisation—a century-long struggle to deal with the cruel social and economic legacy of slavery.

And IBM's mirror world is not the only one that matters in Rio.

Nowhere do the country's contradictions come to a head more than in the fragile squatter settlements that cling to the hillsides above Rio's posh neighborhoods. For more than a decade, along the boundary that separates the Pereirão favela from the surrounding forest, a group of boys have painstakingly constructed an elaborate scale model of their community, cobbled together from cinder blocks and LEGOs and the very mud upon which their own neighborhood stands. Alessandro Angelini, doctoral student in anthropology at the City University of New York, has studied the boys and Projeto Morrinho, as they call their model, for several years.

Much like the mirror world in the Rio Operations Center, the boys' model provides a kind of topsight, a view of the workings of the favela as a whole. But, it is also a stage for acting out the everyday stories of the street using LEGO avatars as actors—stories that provide insight into why the people who live there act the way they do. Angelini's films of their performances run the gamut from *Stand By Me*–style boyhood epics to wild strobe-drenched scenes of the infamous baile funk street parties, where local drug lords tote assault rifles on the dance floor. Whereas IBM's model senses from a distance, the boys' model is driven by observations on the ground. It's a rich reflection of the social gyrations of the favela that are hidden even from the government's view. It is their own representation of the "chaotic multi-sensual reality" that Gelernter saw as the essence of the romantic view of the world, and the side of humanity that mirror worlds would edit out.[96]

IBM's creation encodes the entire city into an inelastic stream of data, but the boys' spins an enriching oral history of a typical favela's human journey. The computer model may tell us what is happening, but the boys' tells us why. The boys' approach is undoubtedly the way any community would prefer to be modeled, not as a collection of objective physical measurements but as the subjective story of a living, feeling organism.

Angelini has a photo of Projeto Morrinho that shows a tiny replica of a real billboard located nearby, which the boys have placed overlooking the miniature favela from on high. "God knows everything but is not a snitch," it reads. While it is merely an ad for a 2008 documentary made about the boys, it's an unwitting reference to the silent watchers in the Intelligent Operations Center. It's as if the boys' mirror world senses the technocrats out there as well, reducing the city, and their very lives, to a set of equations, approximations, and data points.

3

Cities of Tomorrow

In the 1850s, as Ildefons Cerdà envisioned a new Barcelona, he wasn't on the railroad or the telegraph company's payroll. He was merely trying to craft a better city by exploiting new technologies. But today big technology companies have usurped a leading role in shaping our visions for future cities.

These new technicians aim to harness the technologies of ubiquitous computing and a new scientific understanding of cities to transform how we manage them. As we have seen, this isn't the first time technology has played a starring role in the story of urbanization. The massive cities of the Industrial Revolution depended as much on advances in information processing and communications as they did on the rise of steam-powered machines and electricity. In the twentieth century we continued to repeatedly reshape our cities to accommodate and exploit new technologies, wielding new scientific ideas to justify and speed their spread. But employing science and technology in service of reshaping cities has often led to more sorrow than success. We are not the first generation to turn new tools to the problems of cities. But are we clever enough to learn from past mistakes to do it right this time?

From Garden City to Conurbation

By the end of the 1800s, the governments of Europe and the United States faced an urban crisis as dire as the one China, India, and Africa do today. The poor were crowding into booming cities faster than the physical and social infrastructure could be expanded to serve them. There was too much pollution and crime and too little housing, education, and health care. In London, where millions lived in penury, responses ran the gamut. The ruling elite simply abandoned its toxic core for the countryside. Some reformers stayed behind to create new social institutions to help feed, house, and educate the worst off.

Still others argued that cities themselves were the root of the problem. Ebenezer Howard, a clerk for the British Parliament, proposed a simple solution. Start over. A self-made utopian, in 1871 Howard had traveled to America at the age of twenty-one to try his hand at farming in Nebraska. But he was soon drawn to Chicago, where he worked as a shorthand reporter for several years. The city was hastily rebuilding from a devastating fire, largely along its existing lines. Howard watched as a golden opportunity to improve the city was squandered. (Not until Daniel Burnham's ambitious 1909 plan would Chicago articulate a more modern design and lay out the city's majestic public spaces we know today.)

After returning to England in 1876, Howard grew increasingly frustrated with the inability of government to tackle the rapidly worsening problems of cities. By 1898 he was finally ready to propose a more rational approach to city planning and design in the only book he would ever write, *To-Morrow: A Peaceful Path to Real Reform*. In 1902, this manifesto was republished as the deliciously Victorian sci-fi tome planning aficionados around the world now know simply as *Garden Cities of To-Morrow*.

Today, computers provide the technological metaphor that defines our visions of smart cities. Howard drew on the new science of his day—electromagnetism—to describe his model of society. The city and country, he argued, acted as opposing "magnets," each attracting and repelling people through different innate characteristics. Cities and towns offered jobs and opportunities for social interaction, while the countryside had fresh air and cheap land. The city's pollution and high rents pushed people away, but so did the boredom of rural life.

The Garden City, Howard proposed, would be a third magnet, a new kind of settlement that combined the most attractive elements of both city and country. Leafing through his plan for utopia, it's clear that much of his design didn't survive its encounter with car-obsessed America. With its town center and dense bands of multifamily housing, the Garden City looks less like exurban sprawl and more like New Urbanism, the design movement that swept across America in the 1990s with its emphasis on walkable neighborhoods. But many of Howard's ideas, such as relegating industry to the city's outskirts and clustering shops in a massive covered complex at its center (e.g., a shopping mall), are fundamental motifs in American suburbia.[1]

The Garden City was the Songdo of its day—network technology undergirded its daring break from the past. While Londoners choked on smoke from a million coal-fired furnaces, Howard's utopia would run on clean municipal electricity (which, as we saw in chapter 1, had made its world debut only recently in London's suburbs in 1881). More importantly, *Garden Cities* galvanized a growing movement of architects, engineers, and social reformers around rational, comprehensive approaches to the problems of the city. Universities quickly formed programs to train city planners, and by World War II, a whole new profession had emerged. Its practitioners brought Garden City–inspired communities to life throughout Europe and the United States. In 1939, the Regional Planning Association of America, their national organization in the United States, produced a film that captured the excitement surrounding the scientifically designed, technologically powered transformation of the nation. Screened at the same World's Fair in New York that featured General Motors' Futurama exhibit, the film heralded a vision directly descended from the Garden City. "We see homes with grass, children riding bicycles, and men walking to work in clean factories and playing softball," recount historians Robert Kargon and Arthur Molella. It prefigured today's smart city ambitions. "The world of mankind and technology is in balance once again. The lost Eden is restored by good sense, good planning, and good technology."[2]

Garden Cities set the stage for twentieth-century suburbanization. But Howard's design might never have caught the public imagination were it not for the help of Patrick Geddes, a polymathic Scottish biologist turned social planner. Howard sought to work from a clean slate, but Geddes believed that mass urbanization was not to be feared. "Civics," as Geddes called the application of the then-new field of sociology to practical problems, intended to address social decay by mending the physical structure of existing cities. In stark contrast to utopian designers like Howard, who took a decidedly paternalistic approach to the problems of cities, Geddes believed that progress required the full participation of every citizen. A utopian design, no matter how effective, was insufficient. "Whereas Howard proposes a plan," Kargon and Molella argue, "Geddes announces a movement. Howard, the utopian, lays out a map within which change would arrive, but Geddes elaborates a vision of citizenship ('civics') that will prepare a population to build its change."[3]

Trained as an evolutionary biologist, Geddes saw the city as an organism rather than a machine, in stark contrast to the engineers and architects who dominated the nascent urban planning movement. "Forms of life and their emergence and development in interaction with the environment were to become a major interest of Geddes," writes biographer Volker Welter, "determining his life work from his earliest publication to his last book."[4] This unique perspective bestowed Geddes with a view of cities and their evolution that was vast and comprehensive in scope, and he was determined to use it to resolve the growing conflict between city and countryside that Howard's design had sidestepped. "It takes the whole region to make the city," he wrote. City and country were simply different parts of the same biological system. Building on his early work in biological classification, Geddes developed a research method he called the "regional survey," designed to capture a comprehensive snapshot of the entire scope of human settlements, from center to hinterlands. It was also a tool to map their evolution in history. "A city is more than a place in space," Geddes declared to a group of planning enthusiasts gathered at the University of London in 1904, "it is a drama in time."[5]

But Geddes also believed that citizens had "forgotten most of the history of their own city," as he wrote in his 1915 book *Cities in Evolution*. If they were to rally behind a progressive, organic, and scientific approach to city planning, they needed to relearn that history. In 1892 he set out to teach them, putting on display a massive regional survey of Edinburgh. Housed inside an old astronomical observatory in central Edinburgh that Geddes renamed Outlook Tower, it was an immersion center for civic education. Starting on the roof, visitors began by taking in a sweeping live view of the region, presented inside a camera obscura—a kind of room-sized pinhole camera. As they descended from the roof, they passed through a succession of chambers that portrayed the city situated at ever-larger scales—within Scotland, within Europe, and in the world—a Victorian precursor of sorts to Rio de Janeiro's digital dashboard. The building doubled as a repository for the vast archive of information Geddes had gathered about the region, which he intended visitors to experience in its entirety. Upon reaching the ground floor, visitors were ushered out the door into the real city itself.

The Garden City movement spread quickly in the early decades of the twentieth century, its principles inspiring copycat designs around the world. But while Geddes would go on to create several city master plans himself, including Tel Aviv and dozens of Indian cities and towns, it was Howard's precise physical program that attracted the most attention, from fans and critics alike. Jane Jacobs excoriated Howard in *Death and Life of Great American Cities*, published in 1961, arguing that "He conceived of good planning as a series of static acts; in each case the plan must anticipate all that is needed. . . . He was uninterested in the aspects of the city which could not be abstracted to serve his Utopia."[6] She showed little love for Geddes's legacy, the regional planning movement, either, heaping scorn on urban historian Lewis Mumford, Geddes's most influential and loyal disciple in America. But ignorant of Geddes's insistence on full citizen participation in city building, Jacobs's own work reinvented the ambitions of the Outlook Tower. Her book was itself a regional survey of sorts—a carefully studied and holistic dissection of the social ecology of urban life, delivered in plain prose to a huge public audience. And her critique of top-down planning was entirely consistent with the evolutionary biologist's understanding of cities. As historian Robert Fishman summarized Jacobs's argument, the planning elite "completely failed to understand and respect the far more complex order that healthy cities already embodied. This complex order—what she calls 'close-grained diversity'—was the result not of big plans but of all the little plans of ordinary people that alone can generate the diversity that is the true glory of a great city."[7] Geddes would have been proud.

Jacobs so thoroughly skewered Howard's top-down utopian approach that it is still forbidden territory for city planners today (at least in the West).[8] There was much to criticize. The physical master planners who followed in the steps of Howard overreached, destroying vibrant neighborhoods and virgin farmland to make way for lifeless megaprojects. As Tom Campanella puts it, "Postwar urban planners . . . abetted some of the most egregious acts of urban vandalism in American history."[9] The Garden City dream has metamorphosed into the banal reality of suburban sprawl. Another Geddes neologism best describes that unbroken patchwork of built-up areas we now inhabit—"conurbation."

Car Wars

The men—for they were almost all men—who followed in the footsteps of Ebenezer Howard intended to clear-cut slums and countryside alike to make way for progress. They sought to solve the problems of the city by changing its shape, and counted on new technology to stitch their new designs together. But even as they reorganized neighborhoods and regions around the potential of trains, telegraphs, and electrical grids, another technology was emerging whose impact on the physical form of cities would dwarf them all. And in the wake of its devastating impacts, would fundamentally transform the way we plan cities as well.

It all began in Detroit, with Henry Ford's masterpiece of manufacturing management, the assembly line. Until then a luxury good, almost overnight automobiles became a mass-market product. They took American cities by storm. Today we think of New York City as a place where one can escape auto-dependency and walk or take transit instead. But in the 1920s it was a hotbed of enthusiasm for this new means of locomotion. During that decade, the number of registered motor vehicles almost tripled, from 223,143 in 1920 to nearly 675,000 in 1928. The crowding of so many cars and trucks into the densely populated metropolis paralyzed city streets. "A Rising Tide of Traffic Rolls Over New York: What Is Being Done to Relieve the Ever-Growing Street Congestion Which Threatens To Slow Up the Vital Processes of Life in the Metropolis," screamed a *New York Times* feature headline in February 1930. The newspaper projected some 1.2 million motor vehicles would overwhelm city streets by 1935.[10]

Throughout the United States, the arrival of huge numbers of cars and trucks in densely populated cities sparked violent conflicts, pitting pedestrians against a newly motorized elite. The battle was literally waged in blood in the streets. Today, most deaths caused by automobiles occur on highways and in rural areas, and most urban accidents are low-speed and nonfatal. But in the 1920s automobiles plowed through city crowds like juggernauts. The vast majority of the deaths in the early days of motorization were urban pedestrians. "After World War I, the scale of death and dismemberment on roads and streets in America grew fast," writes Peter Norton in *Fighting Traffic*, his fascinating history of the period. "In the first four years after Armistice Day more Americans were killed in automobile accidents than had died in battle in France. This fact was widely publicized, and the news was greeted with shock."[11] Cars and trucks killed some fifteen thousand people annually in the early 1920s—in New York City, there were some thirteen hundred traffic fatalities in 1929 alone.[12] Mob lynchings of offending drivers were common.[13] Children bore the brunt of the attack, mown down at play in streets hitherto considered their domain. In 1925 one in every three victims of the automobile was a child. That year, cars and trucks killed seven thousand children in the United States.[14]

The battle for America's streets lasted less than fifteen years. By the second half of the 1930s, the automobile had clearly won. A massive public awareness campaign orchestrated by newspapers, community activists, and public officials had hammered home the safety risks of jaywalking and allowing children to play in streets. But it was car enthusiasts who dictated the future shape of American cities by enlisting the growing cadre of professional traffic engineers who advanced a new science of street design by appealing to two broad new ideals—efficiency and modernization. Before the widespread introduction of traffic signals, the influx of cars into American cities created the same kinds of hellish traffic jams we see today in Bangkok or Lagos. Applying scientific methods to understand and design systems to reduce congestion offered a quick solution to this new problem. As Norton describes, for the new traffic engineers "streets were public utilities to be regulated in efficiency's name."[15] But when a broad coalition of interests from police to parents to downtown associations mobilized to preserve the status quo, traffic engineers shifted the debate to modernization, painting conventional arrangements around street use as quaint and outdated.[16] They held up the automobile as the ultimate modern ideal—an enabler of freedom and key to the future—a masterstroke of human achievement. Streets would henceforth be reconfigured around the needs and capacities of motor vehicles.

Redesigning the American street quickly evolved into a more expansive project of rethinking the entire national landscape, fueling the transformation of the Garden City concept into modern suburbia. Ford invented the mass-produced car, but it was General Motors that introduced the vision of an entire society organized around the automobile. At the 1939 World's Fair in New York, the streamlined entryway to the company's pavilion (designed by Norman Bel Geddes, no relation to Patrick) transported visitors into a new kind of human settlement made possible by cars. Futurama was a miniature mock-up of a future American city that present-day observers would easily recognize as home. It was an accurate premonition of the cities we've built across the Sun Belt—its sweeping landscape of highways, shopping malls, and suburbs could easily be mistaken for modern Atlanta, Phoenix, or Dallas—a model that China now seems intent on copying en masse. The obvious and intended conclusion of Futurama: new cities must be designed not just to accommodate the automobile, but to exploit its full potential for personal mobility and freedom. In December 1941, with images of Futurama still dancing in their heads, Americans shipped off to war in Europe and the Pacific. When they returned home four years later, they were determined to rebuild their lives according to modern ideals, using every technology at their disposal. At GM's invitation, an entire generation stepped into their cars and simply drove away from the city's problems.

To accommodate the exodus from America's cities, after World War II the focus of traffic engineering shifted to large-scale urban expressways. As Campanella writes, "By then, middle-class Americans were buying cars and moving to the suburbs in record numbers. The urban core was being depopulated. Cities were losing their tax base, buildings were being abandoned, neighborhoods were falling victim to blight."[17] Urban expressways not only gave suburban refugees rapid access to employment in central cities; by allowing the car to take over city streets, traffic engineers' earlier quest for efficiency had already robbed many cities of their once-rich street life. A self-sustaining pattern of decline ensued, as cities emptied out and the car took over.

By the end of the 1950s, organized resistance to urban highway projects had erupted in San Francisco, Boston, and other cities around the country. But it was in New York, where highway construction was displacing hundreds of thousands of residents, that the battle over highways would emasculate not only traffic engineering but the entirety of American urban planning. Robert Moses, the city's planning czar, "was convinced that middle-class families would remain in New York if they could get around by car, and pushed ahead with plans for a comprehensive roadway network for the metropolitan area."[18] Once Moses set his mind to a project, there was almost no stopping him. According to his biographer Robert Caro, he was "unquestionably America's most prolific physical creator." In his long career, he personally conceived and completed public works worth $244 billion in 2012 dollars.[19]

Unstoppable by mayors and governors, Moses the power broker was finally thwarted by a group of Greenwich Village residents. When he proposed in 1952 to extend Fifth Avenue south through Greenwich Village's cherished Washington Square Park, a groundswell of community opposition arose—led mostly by women, including Shirley Hayes, a mother of four, and Jane Jacobs. Throughout the 1950s, the battle waged on as Moses dragged his feet and attempted workarounds such as a depressed roadway with a pedestrian overpass. (A tunnel was deemed too costly.) But by 1958 the tide was turning, and instead of just killing the road, the activists succeeded even in closing the park's existing through roads, a configuration that remains to this day. Moses fumed as he addressed the city's budgetary authority, the Board of Estimate, in a last-ditch effort to save the project. "There is nobody against this. Nobody, nobody, nobody but a bunch of, a bunch of mothers."[20]

Moses resigned as parks commissioner soon after the Washington Square defeat. But the reprisal against Jacobs and company was soon to come. In February 1961, at the behest of James Felt, a Moses protégé and the new head of the City Planning Commission, the city launched a blight study of the West Village, the first step in clearing the way for demolition and redevelopment. As Anthony Flint recounts in *Wrestling with Moses*, Jacobs was dumbstruck when she learned about the plans in the pages of the *New York Times* in February 1961, a month after submitting the manuscript for *Death and Life of Great American Cities*. "Her home and neighborhood, the very neighborhood she had identified as a model of city living in the book she had just written, were now targeted by the urban renewal machine that Robert Moses had set in motion."[21] The blight study was a trick she knew well. "It always began with a study to see if a neighborhood is a slum," Jacobs had noted in her manuscript. "Then they could bulldoze it and it would fall into the hands of developers who could make a lot of money."[22] In place of the funky nineteenth-century neighborhood of bohemians and ethnics would rise modern middle-class tower blocks. Moses envisioned a Garden City in the city. "It was a place to start over, from scratch," Flint observes.[23]

The blight designation was emblematic of the engineering-driven, scientific approach to planning that Howard (and Geddes) had advocated but Moses had perfected and corrupted. As Caro describes, at the headquarters of the Triborough Bridge Authority (renamed the present Triborough Bridge and Tunnel Authority in 1946) on Randall's Island—the most important seat of his power—Moses had assembled an army of draftsmen, engineers, and analysts to survey, document, and design. Moses always had plans at the ready long before legislatures got around to funding them. He was the first and greatest practitioner of the "shovel-ready" approach to public works—always have a big project ready to go when a politician needs to make a splash in a re-election campaign. With their superior ability to study the city, physical planners established their authority and defined debates about the city's present and future.

But the residents of the West Village, who couldn't afford a consultant to undertake a survey to challenge the city's blight designation, crowdsourced their own data-driven retort. According to Flint, "The residents volunteered to conduct a study themselves—surveying building owners, residents, and shopkeepers about the conditions of the West Village block by block." The results, compiled by a volunteer who worked as an analyst in the advertising business, showed that the area's housing was not overcrowded, was being well maintained, and provided adequate bathroom and kitchen facilities.[24] The newspapers conducted their own investigations and verified the survey's findings. Pressure mounted, and by the end of 1961, less than a year after she had learned of the blight study, the proposal was shelved. Jacobs had thwarted Moses and the city once again.

Jacobs's battles with Moses were minor skirmishes within the much broader conflict in American civic life during the 1960s, but her efforts cleared the way for the sharply increased demands for citizen participation in city planning and policy making that would follow. Built atop the legacy of paternalistic utopians like Howard, the profession of planning was thrown into crisis. As Campanella recounts, with its underlying assumptions invalidated, the field moved to "disgorge itself of the muscular physical-interventionist focus that had long been planning's métier." It retooled to engage in social planning as much as physical planning. "Drafting tables were tossed for pickets and surveys and spreadsheets," Campanella writes. "Planners sought new alliances in academe, beyond architecture and design—in political science, law, economics, sociology."[25] A new focus on the *process* of planning displaced the primacy of the final outcome, and intended to expand participation.

Planners recast themselves. Previously their role had been that of objective engineers, expected to design an ideal physical solution to be imposed on the city without comment. Now they would serve as expert facilitators of conversations about the future of cities, providing information and analysis that would help communities make their own choices. A new generation of students, radicalized by the broader social struggles of the 1960s, pushed the profession even further, casting themselves as advocates of disadvantaged groups. Since the deck was already stacked against racial minorities, women, and children, the argument went (by developers, corrupt politicians, and planning departments themselves), planners couldn't simply arbitrate between competing interests. They had to mold themselves in the image of civil-rights activists and urban advocates like Jacobs, and become champions of the powerless. By the late 1960s, this intellectual turmoil had paralyzed city planning. The Regional Plan Association of New York produced one of the few big plans of the era in 1968, its Second Plan (the first plan was done in the 1920s). But as Tom Wright, who heads the organization today, explains, the group was so conflicted about the changing role of planning that it merely documented existing conditions—it didn't dare to make any specific recommendations at all.[26]

After a half-century of bigger and bigger plans, we had returned full circle to where Geddes had begun. Geddes championed preservation and surgical redevelopment of existing cities and was strongly opposed to large-scale slum clearance. In 1915 he wrote from India, "The policy of sweeping clearances should be recognised for what I believe it is; one of the most disastrous and pernicious blunders in the chequered history of sanitation."[27] He practiced what he preached. After marrying in 1886, he and his wife had moved into the top-floor flats of an entire tenement block in the James Court neighborhood of Edinburgh. Over the coming years he lived among the poor while orchestrating a dizzying number of renovation projects in the surrounding area.[28] He described this approach as "conservative surgery."[29] As his son Alasdair later recounted, using metaphors from his father's beloved hobby, gardening, "they set about to weed out the worst of the houses that surrounded them, and thus widening the narrow closes into courtyards on which a little sunlight could fall and into which a little air could enter upon the children's new playing spaces and the elders' garden plots."[30] As Mumford described Geddes's approach to revitalization of cities, "he saw both cities and human beings as wholes; he saw the processes of repair, renewal and rebirth as natural phenomena of development . . . "[31] For Geddes himself, the ambition "to write in reality—here with flower and tree, and elsewhere with house and city—it is all the same."[32]

Top-down, or bottom-up? What is the best way to build cities?

Even as Howard and Geddes worked together to advance rational, comprehensive approaches to city planning, their methods were diametrically opposed. City planning still struggles to resolve the discord. Adding to the turmoil, in Western countries, Jacobs's challenge still casts a long shadow over efforts to think big. As Nicolai Ouroussoff, then the *New York Times*'s architecture critic, wrote a week after Jacobs's death in 2006, "the pendulum of opinion has swung so far in favor of Ms. Jacobs that it has distorted the public's understanding of urban planning. As we mourn her death, we may want to mourn a bit for Mr. Moses as well."[33] (Moses died in 1981.)

"How did a profession that roared to life with grand ambitions," wonders Campanella, "become such a mouse?" Jacobs deserved much of the blame. "She was as opposed to new towns as she was to slum clearance—anything that threatened the vitality of traditional urban forms was the enemy. . . . How odd that such a conservative, even reactionary stance would galvanize an entire generation." Worse, the advocacy turn she inspired for a generation of young planners had been co-opted by the NIMBYism of urban elites who "weaponized Jane Jacobs to oppose anything they perceived as threatening the status quo—including projects that would reduce our carbon footprint, create more affordable housing and shelter the homeless."[34]

The car wars show us the awful longevity of the choices we make about technology's role in the city. In the end, despite the social turmoil, the destruction of cities and countryside, the discrediting of city planning, the car remains at the center of the city—not just in America. "In some ways the war is finished," remarked Georges Amar, the head of innovation for the Paris Metro at a New York University lecture in October 2011, "Cars are part of the mobility system." The struggle triggered by motorization produced a more citizen-centric system of planning. But cities paid a huge price. We will continue to pay for those hasty decisions about urban technology for a long time to come.

Meanwhile, in places like Songdo, the Garden City philosophy of starting over is alive and well, and powered by the new network technologies of our era. The rhetoric of technology giants, heralding efficiency above all, is a page out of the traffic engineers' 1920s playbook. At a major summit organized by IBM in 2011, CNN's Fareed Zakaria epitomized this outdated worldview as he shilled for smart cities, declaring, "Everything in your society has to be modernized. Everything has to be smart."[35] Yet, as we have seen, Songdo is setting the pace for much of the rapidly urbanizing world.

By labeling their own visions of cities as "smart," technology giants today paint all others as inferior. But the lessons of the past cannot be ignored. Make the wrong choice in the design of our smart cities, and our descendants may find themselves a century out, wondering what we were thinking today.

Inventing the Internet

The disappointing legacy of the Garden Cities and the battles over motorization are a sobering lesson for those who think they can master-plan smart cities in the coming century. But the way we create new technologies also went through its own grassroots revolution in the twentieth century, which may be just as important in shaping how we design smart cities. Just as the car wars reached their zenith in the 1960s, battle lines were being drawn over another technological system that has transformed the world—the Internet. Its creators faced a similar dilemma over how to design and build it.

The origins and the economic importance of the Internet are part of a much larger debate about the nature of technological innovation and economic growth. The industrial revolution reshaped the material basis of society, introducing technologies and products we still use today. But there are widely differing views on just how that happened. Pessimists like economist Tyler Cowen believe that a handful of breakthrough innovations drove America's economic engine over the last one hundred years. He sees the decline of productivity growth, the pace of improvement in output per unit of input (labor, capital, machinery), in the US economy as a sign that we have finally exhausted the stockpile of the breakthroughs of the late nineteenth and early twentieth century. He writes: "Today . . . apart from the seemingly magical internet, life in broad material terms isn't so different from what it was in 1953. We still drive cars, use refrigerators, and turn on the light switch, even if dimmers are more common these days. The wonders portrayed in *The Jetsons*, the space-age television cartoon from the 1960s, have not come to pass. . . . Life is better and we have more stuff, but the pace of change has slowed down compared to what people saw two or three generations ago." Not only does Cowen argue that big breakthroughs are the true source of technological progress, he doesn't see anything new in the pipeline of the same magnitude. The result, he concludes, is an inevitable "great stagnation."[36]

Where Cowen sees scarcity, Google's chief economist Hal Varian sees abundance. For Varian, the big breakthroughs of the industrial revolution happened only after, and only because of, a new substrate of interoperable technological components that were invented first. In a 2008 interview, he described this process of "combinatorial innovation": "if you look historically, you'll find periods in history where there would be the availability of . . . different component parts that innovators could combine or recombine to create new inventions. In the 1800s, it was interchangeable parts. In 1920, it was electronics. In the 1970s, it was integrated circuits. Now what we see is a period where you have Internet components, where you have software, protocols, languages, and capabilities to combine these component parts in ways that create totally new innovations."[37]

Focusing on the inputs to technology innovation instead of the outputs tells a very different story of how earlier breakthroughs came about, the technological and economic significance of the Internet, and the prospects for a new age of innovation in our own future. For Cowen, the Web (and ubiquitous computing presumably, though he doesn't seem to be aware of it) are merely the last sputters of a technological revolution that began over a century ago. But for Varian, they form the seedbed for potentially rapid, transformative creation via a million tiny steps.

The Internet is a case in point, contrasting these two views on the nature of technological innovation. In the 1970s, telecommunications companies and academic computer scientists battled over the design of the future Internet. Industry engineers backed X.25, a complex scheme for routing data across computer networks. The computer scientists favored a simpler, collaborative, ad hoc approach. As Joi Ito, director of the MIT Media Lab, describes it:

> The battle between X.25 and the Internet was the battle between heavily funded, government backed experts and a loosely organized group of researchers and entrepreneurs. The X.25 people were trying to plan and anticipate every possible problem and application. They developed complex and extremely well-thought-out standards that the largest and most established research labs and companies would render into software and hardware.
>
> The Internet, on the other hand, was being designed and deployed by small groups of researchers following the credo "rough consensus and running code," coined by one of its chief architects, David Clark. Instead of a large inter-governmental agency, the standards of the Internet were stewarded by small organizations, which didn't require permission or authority. It functioned by issuing the humbly named "Request for Comment" or RFCs as the way to propose simple and light-weight standards against which small groups of developers could work on the elements that together became the Internet.[38]

The telecommunications industry saw the design and construction of the next-generation Internet as a big breakthrough. The academics saw it as a combinatorial endeavor.

TCP/IP, the protocol for transmitting data championed by the researchers, won out in the end. Undeniably, we are better off as result. TCP/IP's simplicity allowed all kinds of organizations to implement it quickly. Its openness allowed anyone to connect freely and inexpensively. The ad hoc nature of its ongoing refinement encouraged the best and brightest minds contribute to making it better. But most importantly, freeing itself of the need to anticipate every possible use or flaw, it allowed people to experiment. It's questionable whether the things that make the Internet so valuable today—the Web, Voice over IP, social networks—could have evolved in a network so rigidly defined by the telecommunications industry. The technical, social, and economic evolution of Internet was, Ito argues, a "triumph of distributed innovation over centralized innovation."[39]

Which style of innovation is right for smart cities?

There are aspects of what Cisco, IBM, Siemens, and other technology giants are planning for smart cities that aspire to breakthrough status. They are weaving an array of new technologies—the Internet of Things, predictive analytics, and ubiquitous video communications—into the city on the scale of the electrical grid a century ago. If they succeed in their ambitions, Cowen will be hard-pressed to deny it. But much of what they have done to date is simply cobble together solutions from off-the-shelf components, with little investment in research and development of new core technologies. It is, in a way, the spitting image of combinatorial innovation.

More worryingly, though, the technology giants are out of sync with what we know about how cities need to evolve, at least in part, from the bottom up. They are making choices, about technology, business, and governance, with little or no input from the broader community of technologists, civic leaders, and citizens themselves. That is holding them back. Smart cities could also evolve from the bottom up, if we let them. Both the evolution of the Internet, and the history of city planning, shows us that.

But it is also crucial to recognize that the Internet didn't just emerge out of thin air. The US government played a huge role in kick-starting it. As *Los Angeles Times* columnist Michael Hiltzik wrote, "Private enterprise had no interest in something so visionary and complex, with questionable commercial opportunities. Indeed, the private corporation that then owned monopoly control over America's communications network, AT&T, fought tooth and nail against the ARPANet," the Defense Department's research network that pioneered the technologies that power the Internet.[40] One can find National Science Foundation research grants in the DNA of almost every major advance in the software, hardware, and network designs that power the Internet today.

This is a dilemma that poses some tough choices. Do we try to pick winners and rally our efforts behind a handful of big transformative projects? Some parts of the smart city, such as reengineering the electric power grid, seem to call for Apollo program–scale breakthroughs. Most of the rest is pretty unclear. Should we instead focus on laying the foundations for a diversity of experimentation to unfold, as we did with the Web? Or, if we do both, how do we balance the two and tie them together in productive ways? None of the answers are obvious yet.

We don't yet know how to build a smart city the way we built the Internet. But it's clear from what we now know about the best ways to build cities and create new technologies that we need to start the search for ways to do it.

The Need for Urgent Participation

Patrick Geddes's approach to fixing the problems of cities demanded total participation. This was achievable only by thinking about large-scale transformation as a series of small, incremental changes. Historically, that was the way we always built cities. As writer and architect Bernard Rudofsky explained in *Architecture Without Architects*, traditional cities were designed and built by everyday people, working together as communities to respond to local challenges using local materials. Over long periods of time, they slowly turned the very earth they stood on into buildings of clay, stone, and mud. This "communal architecture" was highly democratized, decentralized, free-flowing, and adaptive.[41]

The creators of the Internet embedded the same kind of thinking in the design of some of our most important technologies. We've all built the Internet together. It is the most participatory construction project in human history. But participation takes time, which is in short supply for those tackling the world's urgent urban problems. Climate change marches on in its complex dance with urbanization—simultaneously cities are (a) global warming's cause, (b) its biggest victim, and (c) our greatest hope for a solution. Health, education, transportation, jobs—all are lacking.

Today, the most progressive cities update their master plan on a five-year cycle. These massive documents are the result of thousands of deliberations and decisions about tough trade-offs. In cities that grew organically over time, those decisions could be made at a very small scale, iteratively, and in response to both local needs and bigger global trends. But as our ability to build has accelerated through improvements in construction engineering, the frenetic business of real estate development, and new financing schemes, that historic way of designing cities has come undone. As a result, in fast-growing cities decisions about the location of different buildings, facilities, or roads have become ad hoc, arbitrary, and ill informed. Architect Rem Koolhaas, who studied the rapid urbanization of China's Pearl River Delta region in the 1990s, described the pace of design there, telling students, "in China, 40-story buildings are designed on Macintoshes in less than a week."[42] One can hardly expect good decisions amid such haste.

Oddly, just as the pace of building the physical world speeds up, there are signs that as computing hits the streets, the pace of innovation is about to slow down, or at least get a lot more complicated. Ubiquitous computing is a thicket of tough design and engineering problems that will take time to sort out. Gene Becker, who launched HP's first forays into ubiquitous computing in the 1990s, argues that stitching computing into the real world is turning out to be trickier than early visionaries had bet on. "Ubicomp is hard," he writes, using the computer scientists' contraction for ubiquitous computing, "understanding people, context, and the world is hard, getting computers to handle everyday situations is hard, and expectations are set way too high. I used to say ubicomp was a ten-year problem; now I'm starting to think that it's really a hundred-year problem."[43] Adam Greenfield, in his book *Everyware: The Dawning Age of Ubiquitous Computing*, goes even further, arguing that, if the goal is the "seamless and intangible application of information processing . . . in perfect conformity with the user's will, we may never quite get there however hard we try."[44] Marc Weiser, the visionary pioneer of ubiquitous computing at Xerox PARC, wrote that compared to the challenge of designing interfaces for the screen, "ubiquitous computing is a very difficult integration of human factors, computer science, engineering, and social sciences."[45] If we are looking to smart cities for urgent solutions, we may need to reset our expectations.

Still, the potential for rapid advances through combinatorial innovation is a tantalizing bet. If the rise of the Internet has shown us anything, it is that organic evolution doesn't have to be slow—though it may be unpredictable. But for a combinatorial approach to smart city technology to succeed, we must quickly move away from the anachronistic visions of Songdo and Rio and engage a much broader universe of ideas, technologies, and innovators. The technology giants' designs are a twenty-first-century upgrade to twentieth-century paternalism, an attempt to solve all of our problems for us. But in doing so, these designs fail to realize the full potential of smart cities.

Technology lifted up city planning in the twentieth century only to help shatter it after a few decades of failed dreams. Planning's long road back to legitimacy and effectiveness has required developing new approaches that involved entire communities in the planning process. Success of any top-down effort to shape the cities of the future will depend on bottom-up participation as well. Geddes lights the way for us. As biographer Helen Meller writes, "His objective in establishing 'civics' was to dispel fear of cities and mass urbanisation, and to release the creative responses of individuals towards solving modern urban problems."[46] Lewis Mumford, who after decades of correspondence (though they only met in person twice) knew him best, said: "What Geddes's outlook and method contribute to the planning of today are precisely the elements that the administrator and bureaucrat, in the interests of economy or efficiency, are tempted to leave out: time, patience, loving care of detail, a watchful inter-relation of past and future, an insistence upon the human scale and the human purpose, above all merely mechanical requirements: finally a willingness to leave an essential part of the process to those who are most intimately connected with it: the ultimate consumers or citizens."[47]

We would do well to follow Geddes's example. A whole cadre of civic hackers is already leading the way.

4

The Open-Source Metropolis

In the fall of 1970, Red Burns picked up a Sony Portapak video camera for the first time. The world's first portable camcorder, the Portapak cost $1,500 (about $9,000 in today's dollars) and weighed nearly twenty pounds. But for Burns, a documentary filmmaker, "it was an epiphanal moment." As she wrote years later, "The skills required to operate the camera were not out of reach for non-professionals. The cost was not prohibitive and for the first time, it was possible for ordinary people to make their own video documents."[1]

Since its launch in 2005, YouTube has revolutionized the way we produce and distribute video. Thanks to the rapid decline in the cost of digital video cameras, for only a few hundred dollars anyone can shoot, edit, and broadcast short films to a potential audience of billions on the Web. Even most phones sold today are miniature studios, with high-definition video cameras and sophisticated editing software included as standard features. But in the 1970s, it was the Portapak and a new urban telecommunications network—cable television—that promised to upend the media industry and transform the way we communicate.

Cable technology was a latecomer to the city, having originally been developed to deliver broadcast television to remote mountain communities. The earliest systems were set up in 1948 in Astoria, Oregon, and Mahoney City, Pennsylvania.[2] While broadcast signals couldn't reach into the valleys where people lived, by placing "community antennas" (or CAs, hence the acronym CATV you'll see on the back of your set-top box) atop a nearby peak, signals could be run by wire down the mountainside to deliver service to nearby homes.[3] But by the 1970s it had become clear that cable's true value was in its much greater bandwidth compared to over-the-air transmissions. Cable could deliver hundreds of channels to the country's big media markets, compared to the dozen or so that served most regions on the VHF and UHF broadcast bands. Investment surged into the construction of cable networks in cities and suburbs, more than $15 billion between 1984 and 1992. It was, according to the industry's trade association, "the largest private construction project since World War II."[4] Today, cable television is so ubiquitous it's difficult to imagine a time when most homes only received a half-dozen channels of programming. But as recently as 1980, the year Ted Turner launched the first twenty-four-hour Cable News Network (CNN), just one in five—16 million of the United States' 80 million households—subscribed to cable.[5]

For video artists like Burns, cable television was YouTube, Facebook, and Netflix rolled into one. She was determined to explore the potential of the new medium. In 1971, a year before Charles Dolan and Gerald Levin launched HBO just a few miles uptown, Burns teamed up with fellow documentarian George Stoney to establish the Alternate Media Center (AMC) at New York University. Where there was once scarcity controlled by big business, cable had created an abundance of distribution channels. Burns wanted to see how communities would use them.

Not far from where those early cable networks first appeared decades earlier, the AMC set up shop in Reading, Pennsylvania. There, with a grant from the National Science Foundation, in 1975 they built a primitive, yet functional, two-way interactive cable television network.[6] Using a split-screen display and telephone lines to transmit voices, the rudimentary Skype-like multiparty video chat room linked three senior citizens' centers. Burns and her team intended to experiment with new ways to deliver social services such as counseling, health care, and education online over television cable links—some forty years before Cisco would craft its own vision of a smart city around interactive video in Songdo in South Korea. Much like today's social networks, the goal was to connect people to each other. "We deliberately set out to use the system as a socializing force," she wrote.[7]

What happened next surprised Burns, who had expected extensive production and training to really get things going. Volunteers immediately filled the new pipes with their own content. One woman created a weekly chat show where she interviewed local politicians and took questions over the voice link from the distributed audiences. Another hosted a chat-room-style discussion that spanned the different locations. Yet another videotaped interviews with the staff at nursing homes, homing in on issues that "were far more relevant to the needs of older people than any questions we might have designed," Burns reported.[8]

As Burns described it to me nearly forty years later, the convergence of amateur video and cable in the 1970s was "a perfect storm." Because cable television was regulated by local governments, the networks had to strike a franchise deal with each municipality where they wanted to operate. And many communities were starting to demand rights to some of the ample array of new channels for "public access" use. Cleverly, Burns teamed up with the cable companies to sweeten the deal and speed the franchising negotiations. Bankrolled by industry and backed by local governments, she launched community video centers in ten American cities. At the centers anyone could learn to shoot, edit, and broadcast their own content.[9] In just a few short years, a growing network of public-access activists had torn down barriers to community broadcasting that had existed for nearly fifty years. They had shown, using the revolutionary network technology of their day, that information and communications technology could empower people in cities. Citizens could shape the technology, and the business and regulatory context into which it would be applied, to meet their own needs.

Cable was only a shadow of the media and communications revolution in store for the 1980s. Sensing what was coming, in 1975 Burns and others at NYU began planning a graduate program that would carry on the work of the Alternate Media Center, training the next generation of media and technology activists. With seed funding from the Markle Foundation, the Interactive Telecommunications Program (ITP) opened its doors at NYU in 1979 with teleconferencing expert Martin Elton at the helm. Urban scholar Mitchell Moss (my own mentor during my master's degree studies in urban planning) stepped in from 1981–1983 and rapidly expanded the program before Burns returned in 1983 to lead it for nearly twenty years.[10]

ITP's ambition was to challenge top-down thinking about technology. "This is an era of technological promise," Burns wrote passionately in 1981. "Not surprisingly, those most invested in exploring the new technologies come from the private sector. The focus of their interest is obvious: cost effectiveness. However, in concentrating . . . on the bottom line, they have neglected the process through which people harness the technology to create a system. That creative process, although difficult to isolate or quantify, is a crucial element in the achievement of that promise."[11] The whole point of ITP, Burns explained to me, was to "stop paying attention to technology, and start paying attention to people."[12]

Burns's assessment of the hopes, ambitions, and potential conflicts that new technologies spurred in the early 1980s was dead on. And as we embark on the development of smart cities, it remains surprisingly accurate and relevant. The technology giants building smart cities are mostly paying attention to technology, not people, mostly focused on cost effectiveness and efficiency, mostly ignoring the creative process of harnessing technology at the grass roots.

But the birth of public-access cable in the 1970s is a reminder that truly disruptive applications of new information technologies have almost always come from the bottom up. Throughout the twentieth century, as broadly useful new technologies have spread, hackers have eagerly adapted them in unpredictable ways. In the 1970s it was portable video cameras and cable TV, today it's smartphones and the Internet. But the basic urge to repurpose technologies designed for one-way communication, like cable, and turn them into interactive conduits for social interaction pops up again and again. Writing in *Rolling Stone* in 1989, just as the cable era was giving way to the Internet, science fiction author William Gibson explained: "The Street finds its own uses for things—uses the manufacturers never imagined. The microcassette recorder, originally intended for on-the-jump executive dictation, becomes the revolutionary medium of magnizdat, allowing the covert spread of suppressed political speeches in Poland and China. The beeper and the cellular telephone become tools in an increasingly competitive market in illicit drugs. Other technological artifacts unexpectedly become means of communication, either through opportunity or necessity."[13] With little to lose, the grass roots readily adapts flexible and abundant new technologies to pressing problems—spreading dissent, eluding law enforcement, or distributing music. When you start paying attention to what people actually do with technology, you find innovation everywhere. The stuff of smart cities—networked, programmable, modular, and increasingly ubiquitous on the streets themselves—may prove the ultimate medium for Gibsonian appropriation. Companies have struggled to make a buck off smart cities so far. But seen from the street level, there are killer apps everywhere.

Today, a nascent movement of civic hackers, artists, and entrepreneurs have begun to find their own uses, and their own designs, for smart-city technology. Not surprisingly, the Interactive Telecommunications Program has become an important center in this nascent revolution. In a sense, its Greenwich Village loft is itself a microcosm of the smart city, a place where a diversity of experience and know-how, infrastructure and technology come together with the challenges of a living city. The result is a flowering of possibility about what smart cities can be, and a radically different approach to imagining them and creating the technologies that will power them. For every hardware and software breakthrough of technology giants, the students and faculty here generate some faster, better, cheaper, and cooler way of doing it. Corporate R&D focuses on efficiency and control in the name of making urban life bearable and economically productive. At ITP, as it is colloquially known, the priorities of this new hacker vanguard are instead about sociability, resilience, serendipity, and delight.

"Too often technology drives an application," Burns once wrote, "because users are intimidated by the technology and do not have a hand in its design."[14] If there's going to be an open-source alternative to the smart city that comes neatly wrapped in a package from Cisco or IBM, it's very likely we'll see it here first.

The City Hack(er)

Walk east from ITP's loft at Broadway and Waverly Place, and a minute or two later you reach the corner of St. Mark's Place. There, Third Avenue—extra wide to accommodate the El trains that ran overhead until the tracks were torn down in the 1950s—is a traffic-filled moat that separates the relatively staid core of Greenwich Village around New York University from the bohemian throng of tenements, head shops, and nightclubs to the east. Students, burnouts, expat Japanese hipsters, and trust-fund kids jostle for space on the narrow sidewalks. A block north, the ghost of punk godfather Joey Ramone still haunts the tenth-floor apartment where he lived out the last days of his life. In the building that once housed the Electric Circus, the nightclub where the Velvet Underground held court in the late 1960s, now resides a chain Mexican joint.

In 2003, across the street in the men's room of the St. Mark's Ale House, I had my first encounter with mobile social software. The wall space above urinals is essential meme circulation infrastructure for

Manhattan's downtown set. With a captive audience, promoters pile sticker upon sticker, which accumulate in a kind of postmodern sediment. On the underside of the toilet, a placement even more clever and impossible to ignore, a sticker reads "dodgeball.com . . . when NYC is your playground . . . now available on the wireless web!" There's a cartoon graphic of a spike-haired kid being beaned in the head by a red rubber ball.

Tracing the origins of the sticker led me to Dennis Crowley, who may just be the first smart-city hacker. In the late 1990s, Crowley had moved to Manhattan to work at Jupiter Communications, a market research firm founded by Josh Harris, one of the most breathless cheerleaders of the Silicon Alley Internet bubble. As a new arrival to New York, Crowley was a heavy user of online city guides. But he thought he could do a better job, and built the first version of a web app he called Dodgeball as an alternative. Today we'd call it crowdsourced. Back then, he described it simply as "a version of Citysearch"—the most popular guide of the day—"[but] you could write your own reviews on it."[15] When Jupiter was acquired by rival Media Metrix as the dot-com bubble burst in the spring of 2000, Crowley was let go. He splurged, spending half of his final paycheck on stickers to promote the service. Dodgeball soon developed a following among the circle of friends he had made at Jupiter, a diaspora of dot-com castaways whom he affectionately calls "kids."

Crowley moved on to a new job at Vindigo, a start-up whose Palm Pilot app was one of the first city guides for a mobile device. Before do-it-all smartphones, PalmPilots—wireless-less handheld computers known as "personal digital assistants"—stood in as digital replacements for paper-based daily planners. This was before 3G, and Wi-Fi was just coming to market, and just beginning its infectious spread. The PalmPilot didn't feature a wireless connection of any kind. Each time you returned to your PC, you snapped the thing into its cradle and hit a button, syncing data across a serial cable. Like other PalmPilot apps, Vindigo used the daily sync as a way of keeping the guide content on your device up to date. But cleverly, it was also a way of soliciting updates and corrections about the real world from the app's users, whom Vindigo recruited to report when someplace went out of business, for instance. For Crowley, it was an adroit solution to the lack of wireless connectivity, and an important lesson in hacking around gaps in the city's still-incomplete digital infrastructure.

After hours, Crowley continued to work on Dodgeball, which was starting to show the serious potential of the social web. By the end of 2000, the site had hundreds of users who had contributed over sixteen hundred reviews of restaurants and bars in Manhattan and four other cities.[16] But it remained a hobby. As Crowley recalls his days at Vindigo, "I was trying to get them to pull social in, but there was just no concept of social at the time."[17] But before he could get anything started, he was laid off once again as the venture sputtered out. He moved to Vermont to work for a winter as a snowboard instructor before returning to New York to enroll in the Interactive Telecommunications Program.

During his first semester in 2002, Crowley built a second, mobile version of Dodgeball. (The one I'd seen advertised in the bar.) In 1999, Sprint had launched the first line of mobile phones with a rudimentary browser for what it called the "Wireless Web." The service was slow to catch on with users because there was not much content available and even the newest phones of the day had tiny displays. But the Wireless Web provided an easier way to experiment with putting content in the hands of users when they actually needed it, as Vindigo had. Where to go for sushi? Best burger? Fancy cocktails? The wireless link to make it work in real time was finally in place.

But Crowley's own technological epiphany lay just ahead. Friendster, the precursor to MySpace and Facebook, launched in March 2003 and news of its digital social circles spread quickly throughout the city's own. "Friendster happened in between our first year and our second year [at ITP]," he recalled in 2011. "I was like, 'Okay, so Friendster has laid the groundwork, so a critical mass of people understand—you have a profile, and you send a friend request, and you collect your friends like baseball cards.' Once you had this idea of the social network, it's like, 'Dodgeball is Friendster but for cell phones.' People understood it."[18]

For the third version of Dodgeball, Crowley wanted to take Friendster's social circles and layer them in real time onto the user-created-venue database he was rapidly accumulating. Friendster eventually fizzled because there wasn't really much to do once you'd collected all your friends, but social networks were a perfect mechanism for filtering the torrent of tip-sharing content being generated by Dodgeball's users. Crowley envisioned a service that combined social networks and tips with the immediacy and intimacy of SMS text messaging, which droves of young people were already using to coordinate social gatherings around the city.

Today, we take for granted the rich ecosystem of software that's available for our mobile phones. But in 2003, building good software for mobile phones was tough for a well-financed start-up, and nearly impossible for a student. Instead of the open Web, wireless carriers exacted tolls for content providers to enter their "walled garden." It was a business model borrowed from the online services of the 1980s like AOL, CompuServe, and Prodigy, who charged steep fees to big publishers for access to their subscribers (in addition to charging subscribers for access to the service). Walled gardens were a sore spot for the industry, setting back the build-out of the mobile web for years. To make matters worse, every wireless carrier used a different set of technologies.

Recalling how Vindigo had worked around the dearth of wireless data, Crowley came up with a hack around walled gardens to build a universal mobile version of Dodgeball. Just as Vindigo had worked around the dearth of wireless data, Crowley found a hack around the walled gardens—e-mail. In 2003, as he set out to build a universal mobile version of Dodgeball, smartphones were still rare. But most new mobile devices could send and receive short text e-mails wirelessly. Recruiting fellow student and dot-com refugee Alex Rainert to the effort, Crowley began building an e-mail-based interface to Dodgeball. After an intense few months of coding, they had pulled together a few thousand squirrely lines of code written in PHP, an open-source computer language for building Web apps. They set it running on an ITP server, where it waited patiently for e-mails from mobile "kids" across the city.

With the e-mail switchboard in place, Crowley and Rainert turned to work on another hack that would provide the glue to turn Dodgeball into a bona fide social network—a new kind of digital behavior they dubbed the "check-in." As Crowley sarcastically describes it, the check-in offered "a way to globally broadcast your location to all of your other laid-off friends."[19] He and Rainert developed a clever coding system to minimize the effort required. Sending an e-mail with "@Tom and jerry" would check you into Tom & Jerry's on Elizabeth Street, the dive around the corner from ITP that became the duo's informal briefing room for reporters and investors. You could also shout a message that would be delivered via wireless e-mail (and later SMS) along with a notice of your check-in to your friends—say, "@Tom and jerry!happy hour is on."[20]

Dodgeball hit the downtown scene like a new drug, and the check-ins started flooding in. The party-prone "kids" Crowley had collected like Friendster friends at Jupiter, Vindigo, and a brief stint at MTV became Dodgeball's most active users. Last night's mayhem became transcribed forever into a database. Blogs told tales of blackout-inducing binges that could only be recalled through a perusal the next morning of the check-in tailings on Dodgeball.com. "Then we got our first blog post on Gizmodo, and then at that point *Newsweek* and *Time Magazine* were looking at the blogs for stories to write," he recalls. Dodgeball spread virally and Crowley and Rainert spun it out of the university as a for-profit venture. From the three hundred or so students and friends who used the service during their grad school days, membership expanded to a thousand at the new start-up's launch. Within a year, over thirty thousand people had logins.[21]

As Dodgeball became a virtual dashboard for a certain slice of Manhattan's digerati, its social graph—the web of friendships recorded in its database, and the flow of check-ins its users created—formed a new kind of urban media that Crowley and Rainert eagerly employed to design new experiences. One tweak tried to help you make new friends. Normally you only saw the check-ins of your direct friends, but if a friend of a friend checked in nearby, you'd get an alert urging you to go say hi. Another experiment turned Dodgeball into a romantic matchmaking machine, letting you declare a "crush" on another user and alerting him or her when you checked in nearby, to give you a shot at a hookup.

Dodgeball was a tantalizingly valuable piece of digital real estate, which Crowley likened to the Marauder's Map from the 1999 best-seller *Harry Potter and the Prisoner of Azkaban*. A magical atlas, the map used little dots to track the location of Potter's fellow students at Hogwarts in real time. When the book's film version debuted in May 2004, it instantly gave Crowley a visual vocabulary to explain Dodgeball's potential to investors. In a short time, word spread to the West Coast, and Google founders Larry Page and Sergey Brin took a shine to it, acquiring the fledgling start-up in May 2005.

Dodgeball's impact on the subsequent development of the mobile web was profound. With nothing but a phone keypad, rudimentary mobile e-mail, and a perfunctory patois of symbols and place-names, Crowley and Rainert inspired other hackers to cobble end runs around walled gardens. Conversely, the service helped pave the way for the app market by showing the wireless industry the huge demand for new software on mobiles. For nascent social networks, it highlighted how important and tricky location would be, but also proposed some creative solutions to the problems that cropped up, including the dreaded "ex-girlfriend problem" (which should be self-explanatory). Dodgeball showed how social software could be with us everywhere, and be fun without being annoying.

Crowley himself is an archetype for smart-city hackers everywhere. Urban economists believe that cities thrive because they create opportunities for people to interact for commerce, learning, and entertainment. But it takes someone who intuitively understands cities to create a *new* way of doing that for the whole world to use. Jane

Jacobs's treatise of good urbanism, *The Death and Life of Great American Cities,* was a love letter to New York City's Greenwich Village, the same neighborhood that both inspired and accommodated Crowley as he conceived Dodgeball. The book glorified how good streets create opportunities for people to meet by chance. Crowley designed Dodgeball as an engine to amplify that serendipitous potential, by constantly prodding us to get up and go make new friends. If she were still alive, how would Jacobs have judged Dodgeball? Trotted out against the best new ideas in urban design, I think the humble check-in might beat them all.

"Pie in the Sky"

By the summer of 2002, another technology was generating buzz among smart-city hackers all over the world, but especially in New York. While Crowley was working on Dodgeball across town, I was busy marshaling a ragtag army of tinkerers, open source believers, and wireless enthusiasts. NYCwireless, as we called ourselves, held its monthly communion on the first Tuesday of every month. The meetings began in the early evening with demos and discussions about new wireless gadgets. They ended, as often as not, well past midnight over beers at a downtown bar. Around tables strewn with empty glasses and bottles, a dozen or more geeks would stay up late making plans to spread free networks throughout the city. Bike messenger bags stuffed with wireless routers, antennas, and patch cables lay underfoot.

One of those nights, I actually ended up in a bar fight wielding nothing but a surplus military laptop. My partner in this crusade to light up Manhattan with public Internet service was Terry Schmidt, an engineer who was fascinated by wireless networks and mobile computing. If I was the community organizer at the heart of this nascent free wireless movement, Schmidt was the mad scientist, pushing the technology to see if it could survive the mean streets of Manhattan.

A month earlier, I had met my weapon of choice for the first time. Schmidt was standing in a light drizzle on Fifth Avenue near the

Flatiron Building. We were on our way to pitch a new wireless hotspot project to a potential sponsor. He beamed as I walked up, and wiped a slick mist off his Panasonic Toughbook's screen with his sleeve. "It's ruggedized," he explained, "milspec . . . rubber gaskets to keep dirt and sand out. Glare-resistant screen. I got it from a liquidator for $400." Tapping into an unsecured hot spot in one of the offices overhead, Schmidt and his city-proof computer were a vision of the future. I held the compact but dense case, feeling like a supporting cast member in some cyberpunk novel. I had to have one. I ordered it that night.

At the bar that night, Schmidt was in my face, shrieking madly, "Let's smash the Toughbooks together! I want to test the cases!" Winding up, we swung the laptops together as others cheered us on like it was some kind of geek grudge match. Much to everyone's surprise, the Toughbooks were truly tough, and survived repeated collisions without shedding any flimsy pieces. As the bartender's shouts to knock it off cut through the fog of war, Schmidt sat down, flipped up his lid, and smiled as his Linux operating system happily ran through its start-up sequence without a glitch. Drinking beer and banging expensive toys around was a fun way to pass the time. But the abuse that Schmidt unleashed onto that laptop was serious business. The fieldwork of lighting up a city one hotspot at a time was going to take a brutal toll, and he wanted dependable tools. The Toughbooks had earned his approval.

Deploying Wi-Fi throughout the world has taken the better part of a decade. Today, almost everywhere you might want to open your laptop and check e-mail, there's a hot spot for you to hop onto. You just assume that the cafe, library, or airport terminal has a wireless connection, although sometimes you might need a passcode or have to pay a modest fee to use it. In the late 1990s, there was growing excitement about mobile computing, but no network infrastructure to support it. Wireless carriers were just starting to build out mobile broadband networks. That would slow to a snail's pace after the telecom industry bubble popped in 2000.

Then Wi-Fi arrived. Its name a marketing trick borrowed from "hi-fi" audio, Wi-Fi was the result of a visionary decision in 1985 by the Federal Communications Commission to free up a tiny portion of the radio spectrum for experimental use without the need for licenses. For years afterwards, those bands were used primarily by garage-door openers and cordless phones because they were prone to interference by stray radiation from microwave ovens. Engineers called these frequencies the "junk spectrum." But by the mid-1990s, a new generation of cheap and powerful digital signal processing chips was under development. They would power advanced radios that could turn junk spectrum into a broadband bonanza. Wi-Fi used this new computational power and a frequency-hopping technique called "spread-spectrum," originally devised for torpedo guidance during World War II by actress and inventor Hedy Lamarr and composer George Antheil, to simply weave its signals around any interference.[22] The result was that computers could now shove almost as much data across the public airwaves as they could over a wire, with no subscription fees. Wireless local area network (WLAN) systems had existed in offices and warehouses for years, but every manufacturer used a different standard. When the universal Wi-Fi standard known as IEEE 802.11b was finalized in 1999, the market coalesced quickly. Apple popularized the technology with consumers through its AirPort line of base stations and receivers, and manufacturing economies of scale kicked in. For a few hundred dollars, you could light up a bubble of connectivity in an afternoon.[23]

Within the laissez-faire wilds of the unlicensed bands, there were still a few rules that severely limited Wi-Fi's usefulness. You couldn't just turn up the signal and blanket a whole neighborhood, for instance. Wi-Fi devices were limited to just one watt of broadcast power, making its range perfectly scaled for the indoor spaces we inhabit every day. Indeed, the standard was designed for these settings. But the faint signals didn't reach far enough to make it useful in outdoor situations. In the suburbs at least, Wi-Fi wouldn't even get your bits to the other side of the parking lot.

The first attempts to hack around Wi-Fi's limited range started on rooftops. History was repeating itself. In the summer of 1901, radio pioneer Lee de Forest had tested one of the first wireless telegraphs on the rooftops of the Lakota Hotel and the Illinois Institute of Technology's Auditorium in Chicago, where he was a professor.[24] A century later, a whole new generation of radio geeks once again climbed ladders to beam bandwidth across cities and towns. Almost as soon as Wi-Fi hit the streets, they developed hacks to concentrate the limited transmission power into focused radio beams that could stretch over longer distances. They replaced the stock omnidirectional antennas, which spread that energy every which way, with directional "sector" and "Yagi" designs that concentrated the signal into a narrower stream like a nozzle on a garden hose. (One homebrew range-extending design, the "Cantenna," could be constructed from $6.45 worth of parts, including an empty can of Pringles chips.[25]) They mounted these arrays on rooftops in San Francisco, Seattle, Portland, and London, and linked them up into wireless backbone networks, communications grids that stretched across entire metropolitan areas, free of airtime charges and independent of the existing telecommunications grid.

In New York, clusters of tall buildings blocked long-distance wireless shots. But NYCwireless had a different use for outdoor Wi-Fi. That same density meant a single low-power Wi-Fi hot spot could cover any one of Manhattan's small but bustling parks and plazas or even a cluster of apartments. After reading in *Salon*, an online magazine, about someone in San Francisco who lit up the bench in front of their favorite cafe, I realized this could be done all over New York City. In the midst of writing my doctoral dissertation on the large-scale geography of the Internet, I turned to approach the problem from the other end. How could we use Wi-Fi to bridge those last few hundred feet from a DSL endpoint to citizens living, working, and playing in the city's public spaces? I posted a note on the website of Seattle Wireless, which had become a central gathering point for would-be wireless communities around the world. Within days

Schmidt and a handful of others had e-mailed me, and we made plans to meet in person.

The first NYCwireless gathering was held in 2001, by sheer coincidence on the left wing's high holy day—May 1, International Workers' Day. Lacking a clubhouse of our own, and with all of us living in tiny Manhattan studio apartments (mine was just 275 square feet), we gathered at a Starbucks on Manhattan's Union Square. In a foreshadowing of his knack for city hacks, Schmidt had launched a crash effort in the preceding week to get NYCwireless's first hot spot up and running in time for our meeting. Using a custom bit he had fabricated especially for the task (and brought to the meeting for show-and-tell), Schmidt had drilled through 18 inches of masonry to string an Ethernet cable from his Upper East Side apartment to a wireless router he lent to the coffee shop in the building next door. As he told a CNN reporter a few weeks later, what motivated his home renovation was common generosity, "I've got more bandwidth than I'm using and I'm willing to share it for free."[26]

From that humble start, over the next year we perfected a guerrilla model for setting up free Wi-Fi: donated equipment, volunteer labor, and a host who would cover bandwidth costs and provide a space for our equipment. We hung wireless routers outside our own apartment windows and on the fronts of local businesses like alt.coffee, a café fronting Tompkins Square Park in the East Village.

Almost immediately, we found ourselves in a digital land rush. As it turns out, we weren't the only ones looking to bring Wi-Fi to the street. But we were the only ones hoping to do so for free. All of the big wireless companies like Verizon and T-Mobile, as well as start-ups such as Boingo, wanted to muscle in and turned our public spaces into a commercial battleground. As we worried that Wi-Fi's wireless commons would be colonized by business, our fears were confirmed when, in December 2002, AT&T, Intel, and IBM teamed up to launch Cometa Networks, a new venture that promised to build a network of 20,000 pay hot spots nationwide. At NYCwireless we shifted strategy, identifying the most important public spaces and "squatting" with our own DIY wireless infrastructure—the idea being that no pay hot spot would make a dime there if a free alternative were already in place. But as industry mobilized we realized that we needed to move beyond guerrilla tactics. We needed more partners that could pay for bandwidth and give us a place to mount our antennas. The big breakthrough came when Marcos Lara, one of NYCwireless's cofounders, picked up the phone and called the people who ran Bryant Park.

Visit midtown Manhattan today, and nestled behind the magnificent Beaux-Arts monolith of the New York Public Library at Forty-second Street and Fifth Avenue, you will find one of the most vibrant public spaces in any city in the world. On a sunny spring day, Bryant Park bustles with office workers lunching and lounging, and in the winter a full-scale ice-skating rink sprouts from the lawn. But in the 1980s, like many of New York's commercial areas, the park had deteriorated into a den of drug dealing and prostitution. Beginning in 1988, the park underwent an extensive renovation headed by the Bryant Park Restoration Corporation (BPRC) that reinvented it as a living room for midtown. BPRC was one of the first business improvement districts formed in New York in the 1980s, a kind of quasi-governmental neighborhood organization funded by commercial property owners to counteract the cutbacks in police patrols and sanitation services during the municipal austerity of the day.

Aside from sanitation and security, many business improvement districts also provide amenities to increase the appeal of their area. Lara pitched Bryant Park's caretakers an ambitious Wi-Fi project that would cover the park's entire 10 acres and turn it into the largest urban hot spot in the world. They had already noticed that in recent years laptops and mobile phones had been allowing people to linger past lunch hour, and they welcomed our offer to volunteer to set up wireless Internet service. Wi-Fi, we argued, would connect the park even more seamlessly to the commercial life of the surrounding business district. Intel provided the wireless equipment—our efforts coincided with the launch of the company's new low-power, Wi-Fi-ready

Centrino processors designed especially for laptops. Bryant Park would provide a unique showcase for the future of connected, mobile computing.

On June 25, 2002, Schmidt flipped the switch and powered up the network's three antennas to bring Bryant Park into the twenty-first century. That summer, some three thousand people would log on, a stunning number at the time because far fewer devices had Wi-Fi capabilities. On Monday nights, when HBO hosted movies on a large screen hoisted at the park's western edge—one of the city's hottest singles' scenes of the time—the network lit up with activity. Just over a year after Schmidt had turned on the first NYCwireless node at his corner cafe, we retired to the park's beer garden to celebrate. He turned to me and grinned. "What's next?" he asked. "What's your pie in the sky?"

I knew we'd hit on a model that could be copied by communities everywhere: volunteer hackers, cheap off-the-shelf wireless equipment, and the support of institutions with an interest in the health of public spaces. I pulled out a map of Manhattan's Financial District, where the Downtown Alliance (another business improvement district) had already hired us to build a hot spot at Bowling Green, the city's oldest park. With Cometa in mind—the company was boasting in the tech press about its plans to place a hot spot within a five-minute walk or drive of every American—I ticked off a half-dozen sites where we could beat them to the punch line.[27] Over the next year, we rolled out seven hot spots across the tiny southern tip of Manhattan, creating the world's first free wireless district.

Those early projects paved the way for city governments to accelerate the spread of public Wi-Fi. Business improvement districts were already seen as a kind of experimental proving ground for new ways of managing cities. If they could do it, many thought, so could a local government. In 2005 Philadelphia launched the municipal wireless movement with a bold announcement of a city-scale wireless project. While Philadelphia's project ultimately failed, as we will see in chapter 7, thousands of communities around the world have successfully built public Wi-Fi networks. Not all of them are free, but even those that aren't have had a major impact attracting talent and tourists and introducing competition into local broadband markets.

Bryant Park became a showroom for sharing our dream of free public Wi-Fi with the world. It was a direct challenge to the telecommunications industry—the massive Verizon corporate headquarters at the corner of Forty-second Street and Sixth Avenue cast its long shadow over the park's western half. More than a decade later, I still meet with visitors from around the world there, to demonstrate firsthand the power of connecting the virtual commons of the Internet and the physical commons of the city center. George Amar, the head of innovation for the Paris Metro, told me that our 2005 meeting there profoundly changed his view of Bryant Park's role in the city's transit system. Disconnected, it was a place for office workers to relax. Connected, it had become a digital waiting room for the massive subway station beneath it.

Today, community wireless groups continue to deploy new hot spots around the world, but their original leaders have moved on to careers and family. The legacy of those heady days lives on and pops up from time to time in the oddest of places. Zuccotti Park, the scene of Occupy Wall Street's encampment in autumn 2011, was one of the original Downtown Alliance hot spots. Though that hot spot was permanently decommissioned during the park's 2005 renovation, protest organizers simply marched up to another one at the 60 Wall Street Atrium to upload video footage. Ironically, that publicly owned private space was housed inside the US headquarters of Deutsche Bank, one of the world's most important financial institutions, which was indirectly financing the hot spot through its dues to the Downtown Alliance.

But by far the most rewarding NYCwireless story is that of Veljo Haamer, who led a successful effort to blanket the Baltic nation of Estonia with free Wi-Fi. A visit to Bryant Park in 2002 inspired him to return home and light up an entire nation with free Wi-Fi. "New York gave me power," he told a reporter in 2011, "and now it's changed Tallinn and Estonia as well."[28]

Pie in the sky, indeed.

Citizen Microcontrol

At the grass roots, the life cycle of technology innovation is now measured in months. By 2005 Crowley and Rainert had sold Dodgeball to Google and set up shop in the search giant's New York office. But just as Crowley had struggled to get Vindigo to pay attention to social software, Google was slow to see Dodgeball's potential as well. The service languished for years until Google decided to finally pull the plug in March 2009.[29] Meanwhile, the community wireless movement had quietly faded away as municipalities began to take over the deployment of public Wi-Fi access on a larger scale. But back at the Interactive Telecommunications Program a new effort in city hacking was spinning up and cracking open the black arts of sensing and actuation in a direct challenge to industry's vision of the Internet of Things.

One can imagine that hanging on a cubicle wall at Cisco or IBM there's a list of the world's priorities for connecting things to the Internet. If they are even on the list at all, I suspect houseplants are close to the bottom. But if we think about the most basic of human physiological needs—oxygen—the value of a tweeting ficus tree is obvious. Unless you've got a green thumb, however, keeping those symbiotic companions alive can be a challenge. It should come as no surprise, then, that a bunch of students would try to crowdsource it.

"Today's plants are abused, neglected and misunderstood," explains the professorial narrator of the Botanicalls project's retro-1950s promo reel.[30] "Modern life and an increasingly technological and automated society leaves little room for our leafy green friend, the plant." The long-term survival prospects for plants living amid a community of busy grad students were even slimmer still, and so

Botanicalls, developed for a 2006 class on sustainability, turned social networks and the Internet of Things to the challenge of gardening. It was an elegant and simple hack, leveraging a modicum of technology to organize a shift in group behavior. First, the students connected a tiny computer to a moisture sensor wedged among the plant's roots and tethered it to the Internet via a network adapter. As the moisture readings were pushed up to a Web server in the cloud, software designed by the students analyzed the data, triggering a cry for help when it detected dryness. Hooked up to Twitter and the phone system, the contraption let "plants call for human help."[31] The plant's "friends" could follow its Twitter stream to keep tabs on its requests for water, exchange messages among themselves to coordinate care, and receive its expressions of gratitude when its thirst was slaked.

As clever as Botanicalls was, what's most remarkable is how easy it was to bring to life. Just a few years earlier, building a networked sensor would have meant building a circuit from scratch. Instead of making funny videos to promote their invention, students would have spent their evenings holding smoking soldering irons, staring bleary-eyed into a tangle of wires. But Botanicalls is just one of thousands of projects that are exploiting a new approach to prototyping networked objects, allowing civic hackers, students, and artists around the world to invent their own visions of the Internet of Things.

Botanicalls, like many objects on the Internet of Things, is powered by an unsung but utterly ubiquitous kind of computer called a microcontroller. Microcontrollers are the brains of the modern mechanical world, governing the operations of everything from elevators to the remote control on your TV. Like a personal computer, they contain a processor, memory, and input/output systems. But unlike PCs, microcontrollers are small, simple, and cheap. They aren't general-purpose machines that can run a word processor as easily as they play a game—they are optimized to perform just a few functions but do them well, over and over, without crashing. Sensors that measure light, sound, or—in the case of Botanicalls, moisture—

trigger their maneuvers. Preloaded code on the microcontroller analyzes those measurements, determines an appropriate response, and then relays instructions to another add-on. A PC outputs to a screen or a printer. A microcontroller outputs to other devices that act on the physical world—motors, lights, and relays.

Faculty at the Interactive Telecommunications Program began to experiment with microcontrollers in the 1990s to create interactive artworks. In 1999, Daniel Rozin assembled a stunning mosaic "mirror" of 830 tiny wooden tiles, each manipulated by its own microcontroller. Paired up with a video camera focused on the viewer, the motors would deflect the tiles to create different shadings.[32] The result was a constantly changing pixellated self-portrait reminiscent of the work of painter Chuck Close. But at the time, working with microcontrollers required navigating a steep learning curve. Microcontrollers were general-purpose industrial components, designed to be a starting point for electrical engineers to devise complex circuits, not a plaything for artists.

By 2004, two other ITP instructors, Dan O'Sullivan and Tom Igoe, had amassed enough experience tinkering and teaching with microcontrollers to write an introductory textbook for would-be hardware hackers, *Physical Computing: Sensing and Controlling the Physical World with Computers*. But the microcontrollers available to hobbyists and hackers, such as the PIC (Peripheral Interface Controller), were hardly plug-and-play. During a visit to his workshop in 2011, Igoe showed me one, a simple black microchip sporting metal wire legs used to wire it into a circuit board. Sitting on his lab bench, it looked like some kind of silicon insect. "Most microcontrollers are pretty barebones," he laments. "You have to build up a good bit of circuit around them just to get them running. There's no simple software interface for them and you always have a separate piece of hardware that actually flashes the code onto them."[33] What he needed was a cheap and simple microcontroller on which students could quickly load code from their laptops so they could focus on application design, not circuit design. The vast bulk of people interested in physical computing were hackers and artists, not engineers. As Phillip Torrone described it on the blog of *Make* magazine, a kind of latter-day *Popular Science* for hardware hackers, "it's nice to pay your dues and impress others with your massive *Art of Electronics* book, but for everyone else out there, they just want an LED to blink for their Burning Man costume."[34]

The solution to physical computing's steep learning curve came from Italy's own Silicon Valley, the town of Ivrea. Best known as the hometown of pioneering Italian computer maker Olivetti, in the early 2000s Ivrea was the site of a short-lived but highly influential design school, the Interaction Design Institute Ivrea (IDII). Ivrea, like the Interactive Telecommunications Program, was a magnet for hardware tinkerers and attracted students who pioneered improvements on industrial microcontrollers, such as Colombian artist Hernando Barragán, whose *Wiring* prototyping platform was a huge step forward for nonengineers who wanted to experiment with physical computing. For the first time, instead of custom-building circuits around a general-purpose industrial microchip, students could "sketch with hardware," as Igoe put it, incrementally tinkering with sensors, lights, and other actuators. They could also quickly write, debug, and update control code to develop new interactive experiences.

Ivrea shut down in 2005 when new management at its benefactor Telecom Italia cut off funding, but instructors Massimo Banzi and David Cuartielles founded the Arduino project to carry the work forward. The name came directly from a nearby pub, but it was also a clever reference to Arduin of Ivrea, a local nobleman who reigned as king of Italy in the eleventh century.[35] It was also a statement of their aspirations for its role in future physical computing projects, literally meaning "strong friend." As it spun out, Arduino tapped a global community of contributors, including Igoe, who has been one of the project's core contributors. Everything from the hardware on up is open source, allowing anyone to design and manufacture his own variants on the original design.

Today, you can go online to any of a dozen shops and buy an

Arduino that fits in the palm of your hand and does away with much of the labor involved in making a working project with a microcontroller. You can plug it straight into your computer via a USB cable to load your program, and there are a variety of add-on boards, or "shields" (another reference to Arduin), and sensors that can let it see the world around it and connect to the Internet. With Arduino, it is still a few hours' work for the average artist or designer to get that LED to blink. But unlike industrial microcontrollers, the learning curve isn't a vertical brick wall with the instructions for the climbing gear written in an alien language. Once mastered, Arduino can power incredibly complex designs that combine computation and physical objects. "Want to have a Professor X Steampunk wheelchair that speaks and dispenses booze?" *Make*'s Torrone asks. The answer: "Arduino. Want to make a robot that draws on the ground, or rides around in the snow? Arduino." Arduino's magic, he points out, is that it is simple "but not too simple." Amateurs can rapidly prototype new ideas using bits of borrowed code and off-the-shelf components. "It's hot glue, not precision welding," Torrone concludes.[36]

Like any new species of technology, Arduino's real disruptive power lies in its ability to flourish in a new ecosystem. So far, growth doesn't seem to be a problem. When I spoke with Igoe in October 2011, over three hundred thousand officially branded Arduino devices had been sold to date, a number projected to hit five hundred thousand by year's end. We estimated that, including derivative designs and clones, as many as one million Arduinos would soon be "in the wild."[37] Around the world, arts and technology clubs host Arduino workshops to teach the kinds of skills you used to have to go to ITP or Ivrea to learn. RadioShack has even jumped on the bandwagon, and returned to its roots as a hobbyist's supply store during the 2011 holiday season, putting Arduino starter kits and books on display for gift shoppers. Teachers around the world are using Arduino to teach physics and computer science—and blogging about their experiences. Torrone predicts, "Within the next 5 to 10 years, the Arduino will be used in every school to teach electronics and physical computing."[38]

For Igoe, the real potential for cheap, easy-to-use microcontrollers is networking them into clusters that cooperate to create new computational environments. Lean design and mass production have driven the retail price of Arduino boards under $25. While adding a Wi-Fi shield will cost you another $50, prices continue to fall. As Igoe explained, when microcontrollers cost over $100, "you couldn't teach people about computing, you could only teach them about a computer. They would still treat this thing, even though it was cheaper than their laptop, as one computer. Their whole idea, their whole project had to live inside one computer." But as prices fall, more projects incorporate not just "networked objects," as one of Igoe's courses is called, but entire networks of objects. "I wanted [students] to think about computing as a medium. They didn't have to be limited to one central processor. Every object or device could have its own brain, its own processor."[37]

Untethered by Wi-Fi, Arduinos are becoming cheap enough to stick almost anywhere in the city, and could be the raw material for a kudzu-like explosion of a citizen-built infrastructure of urban sensing and actuation. An example is dontflush.me, a system developed by New York City–based designer Leif Percifield. Like many older cities, New York uses a single network of drains for both sewage and rainwater. Normally the combined outflow is processed by treatment plants before being released into the surrounding waterways, but during heavy rains the plants can't keep up; to keep the deluge from backing up into city streets, a nasty mixture of runoff and raw sewage is discharged directly into the city's rivers—some 27 billion gallons a year.[40] But by hooking up an Arduino to a proximity sensor and a $15 cell phone he bought off eBay, Percifield's gadget sits over the outflow pipe and transmits an alert across the Internet to a network of bathroom-based lightbulb overflow-warning indicators.[41] The result is a guerrilla sensor net that encourages people to not flush toilets during overflow events, reducing the discharge of sewage. By changing people's behavior, it could stanch the need for hundreds of millions of dollars of retrofits to the city's sewage infrastructure. Projects like dontflush.me suggest a future where citizens decide what gets connected to the Internet of Things, and why. Instead of being merely a system for remote monitoring and management, as industry visionaries see it today, the Internet of Things could become a platform for local, citizen microcontrol of the physical world.

And that's what's so disruptive about Arduino's growing reach. Torrone suggests more prosaic applications for which Arduino is also the clear technology of choice. "Want to have a coffee pot tweet when the coffee is ready? Arduino. How about getting an alert on your phone when there's physical mail in your mailbox? Arduino."[42] Arduino gives us the tools to thoughtfully structure intelligence into the intimate, everyday, human-scale spaces and objects we live in. It lets us organically wire up millions of tiny wormholes, tubes of code and circuit that shuttle bits and atoms back and forth between cyberspace and the physical world. Instead of big data, it lets us collect and spread a few bits that really matter. The promise is that we'll build the hardware of smart cities just like we built the web, by empowered users one little piece at a time. Botanicalls showed simultaneously how silly but also how incredibly useful and social the Internet of Things could be but, more importantly, it hinted at the creative possibilities that lie ahead.

Don't let Igoe hear you call it an "Internet of Things." It's true that things are being connected and rigged with tiny little electronic brains, eyes, and motors, but for him it is a social technology, a creative catalyst that harkens back to Red Burns's enchantment with portable video, one that lets us pay attention to people instead of technology. Igoe has found that working with Arduino "becomes an excuse to build relationships between people. What happens every time somebody sits down with an Arduino is they turn to ask somebody else for help. Every time somebody makes a new project they'll go and show it to somebody else. They're using it the same way we've used games and other technologies as social lubricant. They get people talking to each other. Right now the problem with the Internet of Things is we get so focused on the thing itself that we fail to recognize that the potential to find new ways to express ourselves to each other through this medium."[43]

As electronics makers all around the world have learned, the most telling sign of success is to have your product knocked off by the "shanzhai" factories of China's Pearl River Delta region just north of Hong Kong. Numbering in the thousands, these tiny, fiercely competitive manufacturers are always looking for a niche to exploit before the others. In 2011, while trying to troubleshoot one student's flaky Arduino, Igoe noticed something was off. The reset button was green, instead of the usual red. Flipping it over, he noticed there was also no Italian flag logo, the Arduino team's patriotic mark of manufacturing quality on the boards. "I asked the student where it came from and she told me she got it at a shop in Beijing," Igoe told me, grinning. "I told her it was a clone."[44]

The shanzhai had voted. If Arduino was worth knocking off, it had truly arrived.

5

Tinkering Toward Utopia

"There is some essential ingredient missing from artificial cities," wrote Christopher Alexander in *Architectural Forum* in the spring of 1965. "When compared with ancient cities that have acquired the patina of life, our modern attempts to create cities artificially are, from a human point of view, entirely unsuccessful." But as much as Alexander revered what he called "natural cities," the appealing ones that had evolved "more or less spontaneously over many, many years," he had little patience for critics like Jane Jacobs who, he argued, "wants the great modern city to be a sort of mixture between Greenwich Village and some Italian hill town." Alexander didn't want to replicate only the appearance of those ancient cities, but rather their DNA. "Too many designers today seem to be yearning for the physical and plastic characteristics of the past. . . . They merely imitate the appearance of the old, its concrete substance: they fail to unearth its inner nature."[1]

Alexander was well equipped to see order in the vast complexity of great cities. Though a professor in the College of Environmental Design at the University of California, Berkeley, he was trained as a mathematician and saw the structure and dynamics of the city through mathematical analogies. To Alexander, the sprawl of postwar suburbia, with its single-use zones and cul de sacs, looked structurally like "trees." In a tree, individual pieces link together up and down in a rigid branching hierarchy, but there are no connections between branches. For Alexander, the architecture and layout of these artificial cities imposed too much top-down order, their individual elements nested like Russian dolls, with each subcomponent enclosed and isolated from those around it.

But "a city is not a tree," Alexander argued in the title of the essay. Cities that develop organically over time possess a rich web of overlapping connections, which to his mathematical brain looked like a semi-lattice. (For simplicity's sake we'll just use the lay term lattice here.) In a lattice, individual elements can be a part of many different sets. They can link up into a hierarchy, or cross-connect in flatter networks.

To explain how lattices worked to create the richness of interactions that he found lacking in modern communities, Alexander described a newspaper rack outside a drugstore near his office in Berkeley. Nominally part of the shop, it became a vital part of the street corner whenever pedestrians waited for the light to change and lingered to peek at the headlines. "This effect makes the newsrack and the traffic light interdependent," he argued. The newsrack, the people, the sidewalk, even the electrical impulses that controlled the traffic signal were woven together in networks of surprising complexity that formed a distinct urban place. Lattices are why the fine-grained hubbub of Greenwich Village or Florence feels so rich and full of wonder, and the single-use suburbs of Los Angeles so empty and banal.

What plagued artificial designs, Alexander argued, was that their hierarchical structure fought against complexity. In theory, because elements in a semi-lattice can be combined with any others, "A tree based on 20 elements can contain at most 19 further subsets of the 20, while a semi-lattice based on the same 20 elements can contain more than one million different subsets." Compare a map of an old, great city with the layout of a modern auto-centric suburb and you will see this clearly. The city is a crisscross of streets and public spaces; there are many ways one could travel across it between any set of two points, interacting with different people, places, and things along the way. But in the suburb, the branching hierarchy of arterials and feeder roads constrains you to a single path. The city is an open grid of possibilities, the suburb a universe of dead ends. "It is this lack of structural complexity, characteristic of trees, which is crippling our conceptions of the city," he wrote. As a remedy, over the next decade Alexander and his colleagues studied traditional cities around the world, distilling their timeless design elements—"the unchanging receptacle in which the changing parts of the system . . . can work together," as he had described the corner in Berkeley.[2] The results, published in 1977 as *A Pattern Language*, were a crib sheet for lattice-friendly city building.

Standing outside the St. Mark's Ale House once again in 2011, almost ten years to the day after I first encountered Dodgeball inside, I browsed the East Village's lattice with my iPhone using Dennis Crowley's newest app, Foursquare. Alexander's ideas about trees, lattices, and patterns have lingered on the margins of architecture and urban design since the 1970s. But they had an enormous impact on computer science, where his writings inspired the development of object-oriented programming. Its philosophy of modular, reusable pieces of code that can be brought together in useful semi-lattices—much like the objects on Alexander's street corner—dominates software design to this day, including the computer language used by iPhone app developers (Objective-C).[3] A fifty-year feedback loop closed as I realized that Alexander's vision of the city as a lattice underpinned the design of the software that now filtered my own view of it.

Foursquare had turned my phone into a handheld scanner that senses the meaningful bits of urban life around me. The home screen opened with a list of nearby attractions: restaurants and bars, shops, even food trucks. A large button at the top urged me to check in, as over one billion others around the world had in the last two years. With Dodgeball, you had to spell out the place you wanted to check in to, and cross your fingers that the system didn't read "Times Square" and mistakenly check you into "Times Square XXX Theater." With Foursquare, putting your pin on the map involves one simple click to select the venue from an automatically populated list of nearby places, and one more to plant your flag.

Digging deeper into the lattice, clicking on people who are checked in at nearby places, I found friends who had recently visited, photos they'd taken, and Twitter-sized tips about things I should do or eat. The app's Radar feature scanned constantly in the background, and chirped an alert about a nearby coffee shop I wanted to check out. It's on a "list" I was following, a scripted guide created by a friend. Lists let you curate collections of places for others to explore—"Best Burgers in NYC" or "Chelsea Art Galleries," for instance. By design, Foursquare was here to do penance for the spontaneity-sapping and serendipity-killing devices of the digital revolution that immerse us in messages from elsewhere as we shamble down the street, oblivious to the world around us. Even more effectively than Dodgeball, Foursquare draped a new digital lattice atop the city's physical one, and connected the two with code. It was perhaps the one piece of software that could turn a skeptical Christopher Alexander into a believer.

In chapter 4 we saw how places like New York University's Interactive Telecommunications Program are generating new designs for technologies that could power more human-centered smart cities. But ITP is just one hub of a grassroots countercurrent of civic hacking, built on open-source and consumer technologies, that is crafting an alternative to the corporate smart cities we toured earlier. Across the globe, others are building on these foundations. In the future, they will create an entirely different kind of smart city, where computers and networks help us connect to each other and the things around us in new and weird but deeply human ways. But can their ideas about smart-city technology grow up and become a real force to be reckoned with?

It had been three years since I last met up with Crowley over a beer right here at the St. Mark's. After ITP, he and Alex Rainert spent two fruitless years trying to convince Google to put resources into scaling up Dodgeball. But the sociable Dodgeball crew didn't fit in at a company where job candidates are screened with math puzzles such as "How many times a day does a clock's hands overlap?" (apparently grammar skills are less prized) or "How many golf balls can fit in a school bus?"[4] When their contract expired in 2007, Rainert went back to Web design and Crowley spent a year on unemployment, wandering around Manhattan's Lower East Side on a used bicycle. Biding time while he waited for the world to come around to his vision, he promised the Dodgeball community that if Google ever abandoned the project, he'd build them a bigger and better replacement. On January 14, 2009, when Google announced it was pulling the plug on Dodgeball, Foursquare was already in the works. That evening, after the first meet-up of a new civic hacker group called DIYcity, I listened as Crowley described the plans he and programmer Naveen Selvadurai had for the new app. It would exploit all of the new technologies that had come on the scene since the early days of Dodgeball.

Two months later, Foursquare launched at the South by Southwest Interactive festival in Austin, Texas, one of the Internet start-up community's biggest and trendiest annual gatherings. It immediately captured the imagination of the tech elite, and after the brief hiatus since Dodgeball had died, a torrent of check-ins flowed once again. Over the next two years Foursquare grew even faster than Twitter or Facebook did in their start-up stage. By August 2011, over 10 million users were collectively logging an average of 3 million check-ins each day.[5] By early 2012, some 1.5 billion check-ins had been recorded worldwide and Foursquare dominated the now booming category of "local, social, mobile" software that Crowley had invented with Dodgeball. Fast followers like Austin-based Gowalla, which had launched at the same festival in 2009 (with a hometown advantage, no less!) failed to keep up. Facebook first tried to buy Foursquare, then competed with its own Places service (with Crowley's former

Interactive Telecommunications Program classmate Michael Sharon at the helm), then bought Gowalla in 2011 and shut it down in March 2012. (All of these moves presaged Facebook's later, more desperate efforts to catch up in mobile apps, such as the $1 billion acquisition of mobile photo app Instagram in 2012.) Celebrities started using Foursquare to promote events and parties. New York mayor Michael Bloomberg, reminiscing about his early days as a tech entrepreneur as he showcased the city's new crop of tech start-ups, visited Foursquare's office on April 16, 2011, to proclaim the city's first official "Foursquare Day" ($16=4^2$). In August 2011, White House staff began checking President Obama in at stump speeches.[6]

I stowed my lattice browser in my pocket and walked over to Foursquare's office on Cooper Square, just upstairs from where the *Village Voice* had chronicled downtown counterculture for over twenty years. (Both organizations would leave the building in 2012—Foursquare decamped a few blocks south to 568 Broadway in SoHo; the *Voice* announced plans to vacate its space to make way for a school). Crowley hadn't strayed far physically or philosophically from ITP, but now instead of hacking together PHP code, he had a war chest of over $70 million, raised from some of the tech industry's most sought-after investors. Out the window, the fast-gentrifying neighborhood pulsed with the creative tension between newcomers and old-timers, rich and poor, hipsters and derelicts. Until 2008, just across the Bowery you could rent a cot at the Salvation Army flophouse for $6 a night. Now you would have to settle for the posh Bowery Hotel just fifty feet to the south, where a suite will run you $600 a night.

It was 10:00 a.m. on a Friday in early May 2011, and a small flock of disheveled twentysomethings trickled into Foursquare's offices with their MacBooks tucked into their bike messenger bags. Tweets and check-in alerts percolated through the air like cricket chirps as the staff slowly recovered from the Foursquare-fueled night before. Being your own lead user is always hard work, but when your product gives you an easy way to find a place to drink and meet new people, it takes its toll. Surrounded by this fast-growing band of coders and designers, Crowley was well on his way to joining the ranks of Mark Zuckerberg of Facebook and Jack Dorsey of Twitter, the princelings of the social web.

On a screen mounted by the elevator, Foursquare's torrent of check-ins unfolded in real time. An animated globe spun slowly, revealing hot spots of check-ins flaring up in a self-service census of the creative class. Berlin, Stockholm, and Amsterdam burned bright as smart young things and their smartphones set out for dinner, drinks, and dancing. With each check-in, they furthered their quest to unlock the app's "badges," a kind of symbolic reward doled out for, say, checking in at four different bars in one night ("Crunked") or at a health club ten times in a month ("Gym Rat"). Crowley came up with the idea after jogging by a spray-painted mushroom ripped from the screen of *Super Mario Bros.* on the Williamsburg Bridge. "Why can't you get power-ups from exploring the city?" he recalled thinking.[7] It's just one of Foursquare's many improvements over Dodgeball.

Why has Foursquare succeeded so wildly where Dodgeball failed? What can it teach us about the ability of grassroots smart technologies to scale up? There are three key ingredients.

First, there was a new, reachable market for mobile apps. The rapid spread of iPhones created enormous demand for new software, and the walled gardens that wireless carriers used to control the Internet experience of users quickly came down. Almost immediately after the iPhone's launch in June 2007, hackers figured out how to "jailbreak" the iPhone's operating system, a technique that allowed them to load third-party software. A little more than a year later, in July 2008, Apple co-opted the growing movement by launching the iTunes App Store. The App Store created a place where buyers and sellers of software for mobile devices could come together and easily do business with a few clicks. While not quite as open as the Web (Apple could and did ban many apps, especially those that replicated the iOS operating system's core features like e-mail), it was a huge improvement.

Second, apps made signing up new users and getting them to interact with the service much easier. Getting on Dodgeball was a complex process—signing up on a website, adding its e-mail or SMS code to your phone's address book, and then tapping out a carefully spelled check-in request to guarantee a match with the system's master atlas of venues. But the App Store could get software into users' hands quickly. You could download Foursquare within seconds of hearing about it from a friend over dinner, and check in before your drinks order arrived. The effect on start-ups was transformative. Once an app caught on, entrepreneurs could take those hard download stats in hand to investors and secure the funds to quickly accelerate development and marketing.

But most important, Foursquare's success was the result of Crowley's experiences with Vindigo and Dodgeball, which gave him a stockpile of ideas to draw from. Both Radar and Explore—another clever function of Foursquare that mines data on your habits as well as your friends' to recommend nearby venues—were things he'd dreamed of building for years. Even as the rest of the industry got hung up on concepts like simply sharing personal location, Crowley was always pushing himself to "do more than just put pins on a map," as he put it.[8] Dodgeball had taught him that knowing where you are wasn't actually that valuable; the value was in using that information to unlock new experiences.

Building on its early success, Foursquare's next move was to become the center of a universe of other apps—a "platform play," in industry lingo. In the years since Dodgeball, the World Wide Web of static documents had evolved into one driven by data, much of it shared and recombined across sites in "mash-ups" of multiple information sources. The Web, like the ancient cities Christopher Alexander idealized, was becoming a lattice of its own. Companies that controlled repositories of valuable data, like Twitter, held a key strategic position. Foursquare had accumulated a similar data stockpile—past check-ins, tips, venue information—but, also like Twitter, couldn't explore every possible use of it. It was time to open up and make itself a piece of the social Web's infrastructure.

Like Twitter and countless other companies, early on Foursquare had launched an application program interface, or API, a structured mechanism that allowed others to write their own apps that would pull data from Foursquare. For example, you could give permission to an app that would repost your Foursquare check-ins to your profile on LinkedIn. The API allowed Foursquare to build an ecosystem of start-ups and hackers that added to the value of its own business but also created thousands of new features that Foursquare either hadn't thought of or didn't see as core features for its millions of users. To seed the community of hackers, Selvadurai hosted "hack days," when Foursquare staff worked with outsiders to build software that plugged into Foursquare. One of my favorites, Donteat.at, built by Max Stoller, a computer science student at NYU, mashed up New York City's health inspection database with your last check-in to warn you off if the restaurant received a failing grade. Crowley must have liked it too, because Foursquare hired Stoller as an intern the following summer.[9]

With its API now used by over 40,000 different apps—many with far more users than Foursquare itself—Crowley's company has established itself as a wholesale provider of location services and data about places for the entire Web and the entire world. It is poised to become a de facto urban operating system, and one that's conceptually light years ahead of anything IBM or Cisco has created. Telemetry and the tracking of stuff—the mere "pins on a map" that Crowley scoffs at—is still the tech giants' killer app for their mundane Internet of Things. For him what matters are the digital breadcrumbs, the pointers that make links between physical and virtual points in the urban lattice. The Foursquare tip by his cofounder, "Naveen recommends the pork sandwich at Porchetta," is more important than where Naveen actually is right now, or even where Porchetta is (110 East Seventh Street). Foursquare doesn't just help mobile, social people figure out where they are. It plugs them deeply into their surroundings in ways we never imagined possible.

Foursquare's success shows how the open, organic structure of the social web's lattice has become a powerful tool for putting people at the center of smart cities. Crowley's meteoric rise has in turn inspired countless tinkerers to turn their own utopian visions of the city into code. But unlike other social-media breakouts like Twitter and Facebook, which were born nearly fully formed (at their core, the basic interaction model of both has changed little since they launched), Foursquare's long incubation shows how hard it can be to engineer the smart city from the bottom up. The technologies are many and hard to plumb together, and interactions between people and the urban lattice are tricky and complicated things to design well. It took the better part of a decade working on Vindigo and Dodgeball before Foursquare's outlines jelled in Crowley's mind. Many of his ambitions still remain unfilled.

Foursquare continues to evolve in response to new lessons taught by its users and the cities they inhabit. In early 2012 Foursquare turned a corner when its users suddenly stopped checking in. As Crowley told TechCrunch, a leading news site tracking the start-up scene, "I asked myself: did we break something? But in fact, it's because people are using Foursquare to look for where their friends are, to find things, and as a recommendation service." Twitter had successfully navigated the shift years earlier, when in 2009 the now-familiar asymmetry of celebrity tweeters to their crowds of followers took shape. "When you start, you are so focused on engagement," Crowley said. "Then you hit this point when you are big enough and say there is something awesome going on anyway. At some point you look and say, oh wow, the consumption model is actually taking off."[10] The first three years of Foursquare was like a massive crowdsourced survey of the world's cities. Now the task was to mine the results and deliver relevant, on-demand recommendations.

What's made Crowley a successful entrepreneur is that he builds things that he would want for himself. But the question remains whether Foursquare can stay true to its roots as it grows into a big company. I'd known Dennis for nearly ten years, watching as his student projects evolved into big business. This was the last time I'd see him face to face for some time, as his responsibilities were growing by the day. I began to wonder if the need to monetize Foursquare was starting to compete with the goal of titillating its users as the realities of taking investors' money started to sink in. Their expectations were high. As Foursquare explored a fourth round of funding in early 2013, its chief backer, venture capitalist Fred Wilson boasted, "Foursquare has more data about real people and the places they go than anybody else."[11]

As my visit in 2011 drew to a close, Crowley had started talking about new features that were in the pipeline. "We're planning . . . this idea of predictive recommendations," he said. He explained how it would work. For instance, if I usually check in around 12:15 p.m., at 11:45 a.m., Foursquare could "ping me with a message telling me where I should go to lunch in the neighborhood that I haven't been to before, but that I might like, based upon where other people have been." It's an experience that, even as a smart-city enthusiast, I'd never considered. I knew I should be excited about it, because I could always opt out, but it gave me the same uneasy feeling I get talking to corporate engineers when they promise to fix cities with big data. I'm not sure I want Foursquare to do that. But I'm sure that marketers and advertisers do.

In the first year after launch, Crowley used to describe Foursquare as a way to make cities easier to use and more interesting to explore. "Check-in. Find your friends. Unlock your city," instructed the company's website. In the beginning, it did that by exposing things out there in the urban lattice we couldn't see directly—our friends, good food, and good times. There was an element of randomness and discovery, like browsing through the stacks at a bookstore. But as data mining and recommendations move to the forefront, Foursquare runs the risk of becoming a quixotic attempt to compute serendipity and spontaneity. The city of Foursquare might look like a lattice, but is it becoming an elaborate tree traced by hidden algorithms? Instead of urging us to explore on our own, will it guide us down a predetermined path based on what we might buy?

The DIY City

For most people the computer age began with the IBM PC, which went on sale in 1981. True geeks, however, date the opening shots of the personal-computer revolution to the launch of the MITS Altair 8800 in 1975. The Altair dramatically democratized access to computing power. At the time, Intel's Intellec-8 computer cost $2,400 in its base configuration (and as much as $10,000 with all the add-ons needed to develop software for it). The Altair used the same Intel 8080 microprocessor and sold as a kit for less than $400. But you had to put the thing together yourself.[12] Hobbyists quickly formed groups like Silicon Valley's Homebrew Computer Club to trade tips, hacks, and parts for these DIY computers. Homebrew was a training camp for innovators like Apple cofounders Steve Jobs and Steve Wozniak who would overthrow IBM's dominance of the computer industry. (According to Wozniak, the Apple I and Apple II were demo'd at Homebrew meetings repeatedly during their development.)[13] Never before had so much computing power been put in the hands of so many.

Grassroots smart-city technologies—mobile apps, community wireless networks, and open-source microcontrollers among them—are following a similar trajectory as the PC: from utopian idea to geek's plaything to mass market. They are being carried along by new communities of civic hackers that share the ideals of the earlier generation of desktop hackers: radically expanding access to technology, open and collaborative design, and the idea that computers can be used for positive change. In 1972, another Silicon Valley hacker group calling itself the People's Computer Company published its first newsletter, with a call to arms emblazoned on the front page. "Computers are mostly used against people instead of for people," it read, and "used to control people instead of to free them. Time to change all that—we need a . . . People's Computer Company."[14] As we have seen, it is a claim that's just as valid a description of smart-city technology today.

And the antidote, once again administered by self-organizing hackers, may be just as potent.

But how does a vague idea about how to use a new technology become a counterculture movement? Sometimes all it needs is a name. John Geraci, another ITP alumnus, is an urban hacker with a knack for naming. In a 2004 class I taught there, on "Wireless Public Spaces," he created Neighbornode, a mash-up of wireless hot spots and community media. Each hot spot hosted a unique local bulletin board that could only be used if you were within range of its signal. But messages could be forwarded by individuals from node to node, in a postmodern game of Telephone. Popular posts could migrate across the city the way a heavily retweeted post on Twitter does today. Neighbornode was cheap and easy, built on open-source software and a $75 Linksys wireless router. As he told the *New York Times*, "If you can install Microsoft Word on your computer, you can set up a community hot spot."[15]

Four years later, John drew inspiration for a new project from a lonely venture capitalist, Fred Wilson of Union Square Ventures. One of the social web's most successful investors, Wilson was also a fan of Shake Shack, restauranteur Danny Meyers's burger stand in Madison Square Park. A sort of spiritual hub for New York's tech start-up scene, the stand was also popular with lots of others, a node in the densely overlapping lattice of Manhattan's Flatiron district. By noon each day, a long line of hungry people stretched an hour's wait along the park's curving pathways.

As one of Twitter's earliest investors, Wilson was always on the lookout for new social hacks to show off the service's usefulness. In 2008 he created a Twitter account called @shakeshack, which people could follow to organize group lunches. More importantly, it was a way to cut the Shake Shack line. As he explained on his popular blog, "only one person has to stand in line and anyone can join as long as they are up for a group lunch with fun people and lively discussion."[16] People soon started sending in reports on the Shake Shack's line to the account. When, less than a week later, local coder Whitney McNamara cobbled together ninety-two lines of Perl code that reposted all of the inbound reports into @shakeshack's timeline, it became one of the first "Twitter bots"—a real-time, crowdsourced ticker of the line's current length.[17]

The Shake Shack Twitterbot showed Geraci that the local Web was quickly moving beyond blogs. After graduation, Geraci had cofounded the first "hyperlocal" news site, outside.in, with author Stephen Johnson. Outside.in brought a geographic sensibility to the blogosphere, aggregating thousands of blogs by neighborhood to create a new kind of virtual newspaper. But the idea of an urban web you could only use from your home or office never seemed quite right. Liberated from the desktop by mobile devices, it could be used to solve real-world problems. Geraci realized that this model had far greater possibilities than just speeding a venture capitalist to his burger.

On October 28, 2008, Geraci launched the DIYcity.org website to convene and challenge the growing band of geeks who wanted to hack their own smart cities. "Our cities today are relics from a time before the Internet," he wrote. "What is needed right now is a new type of city," he continued, perhaps unwittingly echoing the call to arms of the People's Computer Company some four decades earlier, "a city that is like the Internet in its openness, participation, distributed nature and rapid, organic evolution—a city that is not centrally operated, but that is created, operated and improved upon by all—a DIY City."[18] He outlined his vision of an online community where "people from all over the world think about, talk about, and ultimately build tools for making their cities work better with web technologies."[19]

Geraci and I had stayed in touch, and throughout the autumn and early winter of 2008, we would meet for long walks around the East Village, looping out from my apartment at Ninth Street and Third Avenue, on a gallery walk of grassroots smart-city projects. Past the free hot spot a crew of NYCwireless volunteers had installed in early October 2001 to provide relief Internet access after the September 11 terrorist attacks. Past the block where Geraci and fellow ITP student Mohit SantRam had launched the first Neighbornode cluster from SantRam's apartment. Past the café where Crowley and Selvadurai were hard at work coding the first version of Foursquare. As John shared his ideas for building DIYcity, we'd use the city as a brainstorming tool, rehashing the lessons of those earlier efforts.

DIYcity mushroomed overnight. Geraci had built the site using Drupal, an open-source system that allowed anyone to easily form a new group devoted to a specific city or a particular problem. In less than a month local chapters began to organize as far afield as São Paulo, Copenhagen, Portland, and Kuala Lumpur. By the dawn of 2009, thousands of Web developers, urban planners, environmental designers, students, and government employees had enlisted in the effort. With the help of software developer Sean Savage in San Francisco, Geraci organized a bicoastal pair of meet-ups, held on January 14, 2009 (the same day Google announced it would shut down Dodgeball). His goal was to bring together coders and urban planners for the first time to brainstorm an agenda for the nascent movement. Helped along by some free publicity on the popular geek blog BoingBoing, both meetings were packed.

But DIYcity wasn't only about talking. Geraci wanted the movement to build "a suite of tools that residents of any city, anywhere, can plug into and use to make their area better." He had his eye on Washington, DC, where Apps for Democracy, the first city-sponsored apps contest, had run during the preceding autumn. Geraci had concluded that apps contests were an inspired idea but too open-ended and too driven by government data and the programmers' own desires instead of the problems of citizens. So he devised a series of DIYcity Challenges that started with problems—ride sharing, bus tracking, tracking the spread of communicable diseases. To accelerate the process, and keep the focus on users, not tools, he even dictated key parts of the design solution—for instance, a Twitter bot to crowdsource traffic reports. And rather than inviting competition, Geraci's approach was for the entire community to collaborate on a single solution. It was the collaborative culture of Red Burns's Interactive Telecommunications Program reemerging at an opportune moment.

He recruited developers and even worked on the teams himself as they built solutions to the challenges.

The immediate goal was to log a couple of quick wins that showed the DIYcity approach could work. The results were impressive, given the pace of the challenges—which lasted just a few weeks—and the lack of prize money. The first challenge produced DIYtraffic, a service for creating personalized text-message alerts based on a feed of traffic-speed data Yahoo provided at the time, culled from roadway sensors and anonymous tracking of mobile phones by wireless carriers. Presaging the popularity of crowdsourced traffic apps like Waze that would arrive a few years later, DIYtraffic also allowed users to add their own reports to the official feed. In keeping with Geraci's emphasis on reusable tools, a kind of "write once, run anywhere" approach to local software, DIYtraffic was skinnable, meaning that anyone could set up the same service for their own city by simply customizing the outermost layer of the underlying software.

Another challenge focused on public health led to the creation of SickCity, a tool inspired by Google Flu Trends. Both tools sought to map epidemics by mining Internet activity. Flu Trends relied on searches for terms related to flu symptoms and treatments, which Google could geographically tag based on the user's IP address. SickCity was more crude, simply scanning the Twitter stream for keywords like "flu" and "fever." But while it lacked the sophisticated automated methods Google uses to build its list of terms that might indicate illness and had significantly fewer data points, SickCity did have several advantages over Flu Trends. First, people were likely to start reporting symptoms to a social network before the illness was full-blown and they began searching Google for treatments. Second, SickCity offered the ability to see trends at smaller scales—Google didn't start publishing city-level slices until January 2010, almost a year after the release of SickCity. Finally, by changing the filter key words, the tool could be applied to any variety of public health concern, from food poisoning to anxiety.

Created in an all-night marathon of collaborative coding, SickCity was DIYcity's most successful challenge and spread widely in a frenzy of open-source replication. According to Geraci, over one hundred local instances were set up within seventy-two hours. While not scientifically validated like Google's project (a collaboration with the Centers for Disease Control), and flooded with spurious data by the emergence of swine flu (which polluted Twitter with discussions of the disease by people who were not themselves ill), SickCity showed the viral potential of lightweight Web apps that feed off social interactions to address urban problems.

And then, just as fast as it had blown up, DIYcity was gone. After just a single meet-up and five challenges, Geraci made a difficult choice. In 2011, over coffee in Manhattan's Little Italy two years after the end of the DIYcity Challenges, he laughed as he recalled it. "I had a new baby, no job, and wasn't prepared for the success of DIYcity." And as any social entrepreneur will tell you, conceptual success doesn't always translate to financial success. "How do you pay your rent?" he wondered; "It is a question that still hangs over the entire DIY movement, not just DIYcity." As Geraci struggled to find a business model for the project, its early energy was dissipating. Local groups that had formed on the DIYcity site began carrying on their discussions in other forums. "People didn't see a need to stay united," he concluded. Geraci returned to the start-up world. For him DIYcity "lived out its natural cycle. It didn't outlive its usefulness."[20]

But DIYcity did live long enough to become an inspiration, catalyst, and blueprint for organizing civic hacking groups for years to come. It was a People's Computer Company for a generation weaned not on PCs but social media, mobile computers, and open data. It's no coincidence that present in the crowd at that sole DIYcity meet-up in Manhattan was a cadre of civic hackers who would go on to shape the grassroots smart-city movement: Crowley and Selvadurai launched Foursquare a few months later; Nick Grossman and Philip Ashlock of Open Plans would write open-source software for online 311 systems as well as start Civic Commons, a repository for open-source cityware; Nate Gilbertson, a policy advisor to the director of the Metropolitan Transit Agency, would push an open-data initiative through a creaking bureaucracy; and his colleague Sarah Kaufman would see it through.

As Geraci described it, "DIYcity was a totally bottom up organization . . . there was nobody giving orders . . . it was driven by people showing up, looking at what needed to be done, and doing it." Like ITP, "it was loose and collaborative and open and that's what made it work."[21] What Geraci provided was a lens to focus their energy and a well-crafted moniker under which to carry it forward.

Sociability: The Smart City's Killer App

"Use the Internet to get off the Internet," commanded the new marketing slogan for Meetup.com in 2011. Launched in 2002, Meetup was an early pioneer of the hybrid social networks that are commonplace today, bridging online and offline lives to help people congregate face-to-face around shared interests and hobbies. In less than ten years, more than 10 million people had joined over a hundred thousand Meetup groups all over the world. To mark the accomplishment, founder Scott Heiferman reminisced, "I was the kind of person who thought local community doesn't matter much if we've got the internet and TV. The only time I thought about my neighbors was when I hoped they wouldn't bother me. When the towers fell [on September 11, 2001], I found myself talking to more neighbors in the days after 9/11 than ever before."[22]

Meetup's appeal is a powerful reminder that bringing people together for social interaction is the true killer app for smart cities. But we are merely writing the latest chapter in thousands of years of urban evolution—the purpose of cities has always been to facilitate human gatherings. While we celebrate their diversity, as economists such as Harvard University's Ed Glaeser argue, cities are actually social search engines that help like-minded people find each other and do stuff. "People who live in cities can connect with a broader range of friends whose interests are well matched with their own," he argues in his 2010 book *Triumph of the City*.[23] The big buildings we associate with urbanity are merely the support system that facilitates all of those exchanges. As Geoffrey West, a physicist who studies how cities grow, explains, "Cities are the result of clustering of interactions of social networks."[24] And they are repositories of the civilization and culture that grow from these dealings. They are, as urban design theorist Kevin Lynch once put it, "a vast mnemonic system for the retention of group history and ideals."[25] Cities are indeed an efficient way of organizing activity, since infrastructure can be shared. But efficiency isn't why we build cities in the first place. It's more of a convenient side effect of their ability to expedite human contact.

Yet as timeless as urban sociability is, we are experiencing it on a new scale. From the hubs of communication and exchange that sprang up in the markets, palaces, and temples of ancient cities, the size of human settlements has grown, and grown, and grown. Today, the largest megacities tie together tens of millions of people who have come together to work and play in countless groupings. New technologies like Meetup (and Foursquare) are vital to helping people navigate the vast sea of opportunities for social interaction that are available in the modern megacity.

We focus on the physical aspects of cities because they are the most tangible. But telecommunications networks let us see, increasingly in real time, the vital social processes of cities. As much as they enable urban sociability, they are an indispensable tool for studying this ephemeral layer of the city as well.

The telephone has played a key role in urban life for more than a century. Inspired by cybernetics, social scientists first started to study the crucial role of telecommunications in the development of urban social networks in the 1960s, when French geographer Jean Gottmann mapped telephone calling patterns among the cities of the Northeast corridor. In his 1961 treatise *Megalopolis*, Gottmann described how the sprawl of urbanization stretching unbroken literally from Arlington, Massachusetts, to Arlington, Virginia, functioned as a single massive city. In one chapter full of maps he detailed the ebbs and flows of telephone traffic up and down the Eastern Seaboard, arguing that the telephone was the means by which great cities like New York and Washington exerted economic, political, and social dominance over the nation. These cities placed vastly more calls than they received, as their residents gathered information and disseminated decisions from headquarters to the hinterlands. In the 1980s New York University's Mitchell Moss expanded the analysis to the whole world, using similar data to show how Wall Street banks and Midtown media giants were extending this informational trade imbalance to a planetary scale, exploiting new telecommunications technologies to consolidate and dominate entire global markets.[26] In 2008 MIT's SENSEable City Lab brought these studies into the supercomputer age. The "New York Talk Exchange" visualized a year's worth of phone traffic between New York and the world carried over AT&T's global network. On a 3-D rendering of a spinning globe, glowing lines map the flow of calls arcing up from the Big Apple and raining down onto subordinate cities around the world.

It wasn't until very recently that researchers began studying the sociability of cities by looking at the flow of telecommunications happening inside cities rather than between them. In 2006 another SENSEable City Lab project, Real-Time Rome, mapped the movements and communications of an entire city. Drawing on subscriber data harvested from Telecom Italia's mobile network, Real-Time Rome was the first crude EEG of a city's untethered hive mind, depicting millions of fans moving and communicating across the city during Italy's 2006 World Cup victory.[27] As new sources of geographically tagged data from social networks like Twitter and Foursquare proliferate, these diagnostics of urban sociability are becoming more prevalent and more captivating. One of the most compelling projects visualized Twitter traffic in Spain leading up to the massive anti-austerity protests of May 15, 2011. Created by a group of researchers at the University of

Zaragoza, the video is a six-minute snapshot of an entire nation's social network in the throes of a digital seizure.[28]

The sociability of cities isn't all upside. As cities grow, they create social problems too. They typically have higher rates of crime and more disease. But social technology also enhances our ability to address the problems of big urbanism. Nowhere is this clearer than the ways these technologies are created. Whether its Foursquare's API workshops or DIYcity's all-night hackathons, grassroots smart-city hackers all share a vital bit of DNA—the desire to connect, collaborate, and share. They fully leverage the sociability of big cities—the ease of face-to-face meeting, the diverse range of talents and interests—in order to create tools to amplify urban sociability even further. This approach gives them a distinct advantage over big technology companies, where openness is often an impossible cultural mind shift.

Sociability will also provide new tools to address global warming, the greatest threat of all to cities' future. Because cities tend to cluster along coasts, they are especially at risk from rising sea levels caused by the melting of polar ice caps. And so, through organizations like the Large Cities Climate Leadership Group (also known as C40), in the absence of a global compact on climate change, cities from Amsterdam to New York have launched their own coordinated greenhouse-gas-emission reduction efforts. The smart-city visions of the technology industry—increasing efficiency through investments in smart infrastructure—are an important part of these cities' efforts. But efficiency is not enough. Even in Amsterdam, one of the world's leaders, emissions are still climbing.

One promising approach to reducing greenhouse gas emissions that exploits sociability is what design geeks call "product-service systems"—most people just call it "sharing." The basic idea is to use energy-intensive manufactured goods more intensively, so we don't have to make as many in the first place. Take the car-sharing service Zipcar, for instance. By transforming cars from something you own into a service you subscribe to, Zipcar claims that each of its shared vehicles replaces some twenty private ones.[29] Smart technology plays a huge role in making Zipcar practical, by automating many of the traditional tasks involved in renting a car. GPS telemetry tracks vehicle location and use, Web and mobile services eliminate centralized rental depots so cars can be placed close by, and an RFID card identifies allows the renter to unlock one.

But as smart as Zipcar is, it's not very social. But take the same business model and weave in social software to connect people to others with idle vehicles, and suddenly you don't even need Zipcar. San Francisco–based RelayRides helps its members to rent their cars to each other, using a social-reputation system to instill trust and good behavior. While insurance companies have recoiled, three states have passed laws to protect car-sharers from losing coverage.[30] The model is spreading, and now there are social technologies powering peer-to-peer systems for sharing all kinds of expensive private assets. Airbnb does the same for renting out homes for short-term stays, and logged 5 million bookings worldwide in 2011. While they do compete on price with traditional businesses, these services also bait us into more efficient behaviors by turning faceless commercial transactions into human social encounters. It's infinitely more rewarding to rent the poet's flat in San Francisco on Airbnb than to book a soulless hotel room on Expedia.

Sharing systems can be deployed rapidly—often the only additional infrastructure that's needed is the Web. And there are tangible environmental benefits. While spending a night in some hotels is less carbon-intensive than spending a night in the average US home, building the hotel in the first place accounts for a significant share of its total lifetime carbon emissions.[31] Construction is an incredibly wasteful sector of the economy—according to the United Nations' Sustainable Buildings and Climate Initiative, "the construction, renovation and demolition of buildings constitute about 40 per cent of solid waste streams in developed countries."[32] Frank Duffy, an architect who is one of the world's leading experts on workspace design, argued that—at least in developed economies—

we have already built all of the buildings we will ever need. We just need to use them more intensively.[33] Sociability is a strategy for achieving that by motivating us to share; social software now provides the tools to do so widely.

Bugs in the Grass Roots

These new tools are both a better lens through which to see what really makes cities tick, as well as to graft an entirely new latticework for urban sociability onto them. But are civic hackers up to the task of bringing a bottom-up vision of the smart city into existence? Can we evolve the smart city organically—one app, one check-in, one API call, one Arduino, one hot spot at a time? Perhaps, but for all its promise, there are a lot of bugs to be worked out in the grass roots too.

"This is the time for people to throw their hats up in the air and think," Red Burns said with a shrug when I asked her to speculate about the shape of a world filled with all of the mobile, social, sensing things her students at the Interactive Telecommunications Program are cooking up. I guess I'd hoped for some more concrete vision, but she'd nailed the mood of the present. ITP is a microcosm of this movement of young people all across the world who, weaned on the mobile Web and social media, are experimenting with human-centered designs for smart cities. DIYcity was a glimpse of a new utopian vision—open, social, participatory, and extensible—dramatically different than the one technology giants are selling. It wanted to bring into being a smart city modeled not after a mainframe, but the Web.

History is littered with failed plans and false utopias that didn't live up to their promises. Or, as often happens, they evolved in unexpected directions. For Burns, public-access television fell short. "Now I look at public access," she told me, "and I'm disappointed because people don't use it the way I'd hoped." Even ITP turned out differently than she had expected. "I thought it was going to work on social projects like domestic violence. But what happened was when the tools came, people wanted to play."[34] If the risk of corporate visions of the smart city is their singular focus on efficiency, their advantage is clarity of purpose. The organic flexibility of the bottom-up smart city is also its biggest flaw.

Or so say the naysayers. To them, civic hackers are nice kids with good intentions playing with gadgets or trying to strike it rich. City leaders have real problems to solve right now—global warming, decaying infrastructure, and overburdened public services. They don't have time to play with Arduino. They need the might of sustained industrial engineering applied to replumb entire cities over the span of a decade. The grass roots may be a source of new ideas, but what they need is someone who can design and deliver a robust infrastructure that is centrally planned to be safe, efficient, and reliable at a reasonable cost. To an extent, they're right. Scaling up things that work at the grass roots is a challenge few have overcome. Foursquare, even with all its resources, went through a wrenching series of outages before it was able to work out a scalable database scheme (although one of the worst problems was caused by an outage on Amazon's cloud-computing services, the epitome of large-scale smart infrastructure).

Even when they can manage the technical hurdles that come with growth, many civic hacks never get that far. They solve a problem for a small group of users, but fail to sustain the effort to refine their design into something that can connect to a larger audience. As DIYcity's Geraci explained, "it's dead simple to prototype version one of a smart city app. Getting it to version seven, where an entire city's population can use it, is another story."[35] But both scaling and evolving software, it turns out, are exactly the kind of tasks that big companies and professional engineers are particularly good at. Finding ways to effectively integrate industrial engineering and grassroots tinkering is one of the keys to building smart cities well, as we'll see.

More of a problem, though, is the lack of a coherent ideology or even sense of identity. DIYcity was a flash in the pan: there's no equivalent of the People's Computer Company today. And as we've seen, the energy is diffused across different technical communities—wireless geeks, Arduino hackers, apps developers, etc. Their emphasis on openness and collaboration accelerates innovation, but the focus is still exclusively on the technology. There's a growing sense that a "civic tech" movement is coalescing, but it has no clear shared aims.

Even at ITP you can sense this yearning for a larger purpose, for a renewed thrust to complete the unfinished manifesto DIYcity left behind. As Burns showed me to the elevator after a visit in 2011, she tugged on the passing sleeve of John Schimmel, a new faculty member who was building an app called Access Together that would support a crowdsourcing effort to gather data to help disabled people navigate the streets and sidewalks of New York City. Waving a thumb at him, she told me, "This is what I'd do, what I'd work on." I sensed a frustration in Burns that more students weren't driven by the same desire for social impact that drove her as a young woman. While ITP students often have a keen sense of the social dynamics in their tightly knit group (as Crowley did), like anyone engaged in intense study, they often lose sight of the larger world around them. But if this place wasn't going to birth the next People's Computer Company, where could it possibly happen?

Perhaps this new vanguard of smart-city hackers is just navel-gazing kids playing with gadgets. Clustered as they are in the affluent "creative class" districts of New York and San Francisco, should we be surprised when they solve their own problems first? With NYCwireless, it took years before we ventured beyond Manhattan's trendy neighborhoods and refocused on broadband projects in poor areas. Not only do they not represent the full range of the city's people; often these hackers lack a sense that it's even their duty to help others. And unlike the pioneers of the PC and public access cable, they've been raised on a steady diet of personal technology. Hacking is just as often an attempt to seize control of consumer products for personal gain, rather than to employ them in the pursuit of social change. But as the tools to forge a different kind of smart city from the one that industry would spoon-feed us get into the hands of more activists, artists, and designers who yearn for change, will a new social movement emerge?

For Red Burns, the real allure of video was how it democratized visual storytelling. Film was for experts. It needed to be developed and edited, a tricky and time-consuming process that required a lot of training. "But when you work with video, you can see it immediately," she said. "Anyone can learn how to use it, and it throws a whole different cast on communications." It was real time, and that empowered real people. "We would train the students to go into the field to teach people in the communities how to use this equipment and give them the freedom to do what they wanted."

Burns recalled a group that made a video about a dangerous Upper West Side intersection and took it to City Hall to demand a new traffic signal. "They got the light," she said. "I realized it wasn't about technology. It was about community organizing. That, I think, made the difference. I cared about the fact that nobody had a voice."[36]

6

Have Nots

It was hot in Chişinău.

In August 2010 the worst heat wave in a generation baked eastern Europe. Smoke filled the air as wildfires burned across Russia, where the soaring temperatures killed thousands. But in the capital city of Moldova, the most pressing problem was the economy.

A tiny, landlocked backwater of the former USSR, Moldova hides tucked away in the hills between Romania and Ukraine. Once a Florida of sorts for mighty Russia, a coveted retirement destination for Communist Party apparatchiks, it had become the poorest country in Europe. After the breakup of the Soviet Union in 1991, former republics such as Estonia embraced Western-style reforms and thrived. Moldova, however, never managed to shake off Communist influence. After flirtations with democratic reforms in the 1990s, the party was voted back into power in 2001. Over the next decade, the economy imploded, and a quarter of the working age population left in search of work abroad. Twenty years ago Moldova was wealthier than Romania, with which it shares a language and culture. By 2010, when I visited, its per capita GDP was just a quarter of its booming neighbor's.

The previous spring, the country had reached a breaking point. After the Communists narrowly won the April 2009 election in a suspiciously strong showing, outrage turned to violence in the streets. Rallied by investigative journalist Natalia Morar and a handful of social-media mavens, Moldova's "Twitter Revolution" followed the SMS-powered one in neighboring Ukraine a few years earlier.[1] Protestors lit bonfires and waged angry demonstrations in the city center. That June, unable to elect a president, the parliament was dissolved. In the ensuing snap election a coalition of anti-Communist parties snatched a close victory. Within months, they had reached out to the West for help reforming and reinvigorating the economy. At the invitation of the World Bank, I was there to help the new government kick off "e-Transformation," a project intent on leveraging smart technology to modernize the country's archaic bureaucracy. With their uprising, its flames fanned by social media, the Moldovans had already launched their own digital transformation. Our job was merely to help clear the way for it to continue.

I wasn't expecting much from Moldova, starved as it was of talent and investment, both of which had more lucrative prospects elsewhere. The World Bank didn't impress me much either. After decades of trying to slow the growth of cities by investing in rural infrastructure, the organization had only belatedly started to address the planet's new urban reality. Accustomed to start-ups with trendy vowel-deficient names like Flickr and Tumblr, to my ears "e-Transformation" sounded like something from the 1980s. But when I found out that Robert Zoellick, the president of the bank, would travel to Moldova to personally launch the initiative, my antennae perked up.

As deputy secretary of state under George W. Bush in 2005, Zoellick had delivered one of the most fascinating foreign-policy speeches in modern American history, challenging a reluctant China focused on domestic stability to become a "responsible stakeholder" and take a more active role in global affairs. He'd helped mediate the German reunification in the 1990s, and more recently traveled repeatedly to Sudan's Darfur region to intervene in the government-backed genocide occurring there.

Zoellick also was breaking down the World Bank's secretive culture by sharing its data with the outside world. Just a few months earlier, in April 2010, he had announced a new open-data initiative and released online, at no cost, statistics that the bank had long closely held—the World Development Indicators, Africa Development Indicators, and the Millennium Development Goals Indicators (which track progress on the UN's poverty eradication efforts). Soon after the event in Moldova, he would launch a competition to entice programmers to use this data to build apps for development practitioners.

The bank's agenda in Moldova was urgent. With new elections less than a year away, if the country's fledgling liberal democracy was to survive, it needed to deliver reforms and economic results quickly. Failure to meet the electorate's high expectations could send them running back to the familiar, if penurious, stability of Communist rule. But Moldova was also an opportunity to airlift the same ideas about openness that Zoellick was using to reinvent the bank and drop them onto an entire country.

e-Transformation aimed to sweep aside Moldova's entire Soviet-era paper-based bureaucracy and put all government services online. Even in 2010, basic transactions—such as obtaining an exit visa to work overseas—required a long and costly trip to the capital. With $23 million in loans from the World Bank, parceled out over five years, the new government would build a "g-cloud" (a cloud-computing infrastructure that would allow for the delivery of services to both fixed and mobile devices), create a new digital citizen-identity program, and rewrite legislation to encourage private investment in online services. In a country where most rural people still stored their savings under their mattress or in a hole in the backyard, new rules would allow mobile banking. Zoellick spent an hour and a half of his day in Moldova at our workshop (just one of several more conventional programs launched that day). His presence testified to the importance of this project, the first of its kind for the bank and potentially a model for many other countries. On its face, e-Transformation was the worst kind of development aid—driven by an external ideology of neoliberalism, focused on technology, and hastily implemented. But as the people's self-organized Twitter Revolution demonstrated, Moldovans badly wanted change, and saw an important role for mobile technology in securing it. With the bank's help, for better or worse, they were about it get it.

The incongruity between the Communist legacy of privation and the digital abundance of the present was everywhere in Chişinău. In search of a gift for my daughter, I wandered the main street market that comprises nearly the entirety of the city's shopping. All I could find were basic goods—vegetables, drab polyester shirts, school supplies. There was little to the local economy beyond the staples. But just around the corner, a poster advertised 100-megabit-per-second Internet service, delivered to the home over a brand-new citywide fiber-optic network, for the equivalent of $20. Moldova apparently had faster, cheaper broadband than Manhattan or San Francisco. Back home in America, policy makers were wringing their hands over the slow pace of investment in our nation's broadband infrastructure. But here, in tiny, poor Moldova, they'd found a way to make it happen.

The rapid spread of fast connectivity was unleashing the nation's potential. If Moldova's surge into an uncertain digital future was only powered by government, I'd have been more skeptical. But it was also riding a fast-growing wave of entrepreneurship. By 2010, over five hundred technology companies employing some seven thousand people had popped up in Chişinău, little shops of engineers booking over $150 million a year in outsourced work with corporate clients throughout Europe.[2] And that was just the ones that operated in the open. World Bank analysts believed that a parallel shadow industry of freelance web programmers, peddling their services on outsourcing sites like oDesk and Elance and taking payment to offshore accounts, probably generated half that much economic activity again. As wages surged in Russia, the jobs that had popped up there a decade ago were moving south in search of cheaper labor. But it was a passing moment, on which the foundation for higher-value-added industries needed to be swiftly laid. Moldova had only a few years to move up the value chain before Turkey, Uzbekistan, and other places with lower labor costs to the east and south would steal these wage-sensitive jobs away.

It's the fragility of this nascent tech bubble that e-Transformation has to address if the Moldovan experiment is to succeed. The best way to do that, and to ensure the broader success of the country's democratic turn, is to appeal to its diaspora. The hundreds of thousands of bright young Moldovans scattered across the world represent a brain drain of devastating proportions. But thousands are employed abroad in technology firms—over two hundred at Microsoft alone. South Korea, Taiwan, China, and India have all created home-grown tech bubbles by turning the brain drain into "brain circulation," according to AnnaLee Saxenian, who studies immigrant engineers in Silicon Valley.[3] Moldova needs its expats to come home and plug themselves and their social networks back into the local economy. It also doesn't hurt that overseas Moldovans are the country's most strident anti-Communists, and participate actively in the civic life of the country on social sites like Facebook. While they are permitted to vote, they have to go to the embassy in their country of residence to do it. If e-Transformation can bring the polling booth to them directly, the revolution will be secured forever. And, as has happened in India, China, and other countries where emigrants have come home to build businesses, it might just set the stage for their eventual, triumphant return.

ICT4D

My experience in Moldova gnawed at me. Endemic poverty and a host of social ills left little room for debate over the decision to use technology to get results fast, and try to make government work for everybody. Yet the poor were conspicuously absent in the digital utopia put on display by Cisco that same summer at the 2010 Shanghai World Expo. In Eduardo Paes's Rio, they were a problem to be measured and managed with IBM's software, so the Olympic games could go off without a glitch, and the globalization of Brazil could proceed unchecked. The cyber-utopias of the apps start-ups and open-data hackers demanded a college degree, a downtown Manhattan flat, a $400 phone and a crew of hip friends. Everywhere, the people who needed the benefits of technology the most seemed to be missing out or, even worse, suppressed by a new technological elite. Plato's observation in *The Republic* seemed as true in the emerging smart city as when he wrote it more than two thousand years ago, "any city, however small, is in fact divided into two, one the city of the poor, the other of the rich; these are at war with one another."[4]

Around that time, the Rockefeller Foundation came to a similar conclusion—in the rush to wire up smart cities, the poor were at risk of being left behind, or worse. The foundation was already deeply engaged in contemporary urban issues, as it had been since its founding in 1913. Just a year earlier, in 2009, it had published a call to arms for the philanthropic community, the alarmingly titled *Century of the City: No Time to Lose.* The book was a compendium of research presented by a global group of experts during a 2007 workshop at the foundation's Bellagio retreat center near the Italian Alps. It made a compelling case for action, arguing that throughout history, rapid urbanization has always been accompanied by growing inequality and social tension. In the nineteenth century, New York and London packed the poor into tenements in unspeakable living conditions before working them half to death in factories. From the Chinese shadow cities of migrant laborers squatting in abandoned Cold War bunkers underneath Beijing to the Indian and Pakistani guest workers of Dubai who sleep in shipping containers, today's urban boom was rehashing this inequity on an unprecedented scale.

At first, it didn't feel right talking about improving the lot of the poor at the Rockefeller Foundation's headquarters, a $15 million multistory complex of midtown Manhattan office space. An atrium soared over the reception area, where Maya Lin's sculpture *10 Degrees North* provided a serene retreat from the chaotic streets outside, a lavish citadel of granite, wood, bamboo, and cane. But when I visited in the summer of 2010, I found Benjamin de la Peña, the philanthropy's associate director for urban development, to be the physical manifestation of the foundation's new institutional urgency and renewed commitment to cities. In the staccato diction of his native Philippines, he greeted me and ushered me into his office. There were stacks of books about cities and technology, dog-eared and crammed with multicolored sticky notes. A degree in urban planning from Harvard hung on the wall. As he explained to me, de la Peña believed that the destinies of cities and smart technology were now inseparable, but he worried that the explosion of data about cities wasn't only an opportunity for the poor but a huge risk. What did the push for smart cities mean for the projected 3 billion people the United Nations feared would be living in slums by 2050?[5]

There were more questions than answers. What new economic opportunities were there for the poor and other excluded groups? Could city governments use new technology and data to enable e-Transformations of their own? Would the poor suffer from new kinds of victimization at the hands of those wielding tools that could control and exploit them? The challenge, as de la Peña saw it, was to find opportunities for the poor to get ahead or at least keep up, and to shield them from the worst of the unintended consequences. But he needed a map to convince others that there were clear avenues of change that philanthropy could accelerate or try to block. He wanted a forecast of the opportunities and challenges at the intersection of cities, information, and inclusion.

After our meeting I walked the few short blocks over to Bryant Park's free wireless zone, which seemed as good a place as any to draw up a research plan. I had long been interested in the use of technology and data in poor communities. As a college student, I'd run a dial-up electronic bulletin board system out of my apartment and tried (in vain mostly) to sign up kids from the poor parts of town for free e-mail accounts. In grad school, I helped design a wireless network for a public housing project in Boston with fellow student Richard O'Bryant, and later initiated a partnership between

NYCwireless and Community Access, an NGO that builds transitional housing for people coming out of the mental health system. Neither was I a stranger to the challenges of urban poverty, having spent my summer internship in 1994 working for a developer of affordable housing in and around Camden, New Jersey, then the second poorest city in the United States.

Over the preceding decade, a timely confluence of technological change and an international push to end poverty had provided the focal point for a thriving new academic field and activist movement that called itself Information and Communication Technologies for Development (shortened in practice to the slightly less clunky acronym ICT4D). By the late 1990s, as the Internet was powering social and economic transformation in the developed world, people started to think about how its benefits might be exported to developing countries. The pronouncement of the UN's Millennium Development Goals in 2000 brought a renewed international focus on the 3 billion people who at the time lived on less than $2 per day.[6] In the years that followed, thousands of projects were launched to deploy computers and the Internet as tools for education, health care, and economic development in poor communities throughout the world.

By 2008, there was such a huge body of research and activism that Richard Heeks, a professor of development informatics at the University of Manchester, penned a retrospective look at what had clearly become a movement. "ICT4D 1.0," as Heeks described that first wave of efforts, had largely been a reckless failure:

> With timescales short and pressure to show tangible delivery, the development actors involved with ICT4D did what everyone does in such circumstances: They sought a quick, off-the-shelf solution that could be replicated in developing countries' poor communities.
>
> Given that poverty concentrates in rural areas, the model that fell into everyone's lap was the rural telecottage or telecenter that had been rolled out in the European and North Amer-

> ican periphery during the 1980s and early 1990s. Understood to mean a room or building with one or more Internet-connected PCs, this model could be installed fairly quickly; provide tangible evidence of achievement; deliver information, communication, and services to poor communities; and provide sales for the ICT companies that were partners in most ICT4D forums. Thus, a host of colorfully named projects began rolling out, from InforCauca in Colombia to CLICs in Mali to Gyandoot in India.[7]

With greater effort going into marketing and publicity than end-user engagement and financial management, few of the telecenters were sustainable. "Sadly," Heeks continued, "these efforts often resulted in failure, restriction, and anecdote."

The most stunning telecenter failure was the work of one of the world's most revered technology academies, the MIT Media Lab. Little Intelligent Communities (Lincos) was a brilliant design that packed what it called a "digital town center" into a shipping container that was connected to the Internet by satellite.[8] The idea was to air-drop the boxes into remote villages, thus plugging them into a global web of learning, culture, and commerce. In 2000 the first Lincos telecenter was installed in the Costa Rican town of San Marcos de Tarrazú. But just two and a half years after it opened, its initial operating subsidy exhausted, it shut down. Only one other telecenter was installed in Costa Rica, at the Costa Rica Institute of Technology, to be used as a monitoring center for a proposed nationwide network of Lincos sites that was never to be.

Subsequent efforts to scale up Lincos in the Dominican Republic showed that the novel containerized design itself was also deeply flawed, not just its subsidy-hungry financing scheme. The plan for that country called for sixty Lincos boxes scattered around the countryside, a number quickly reduced to thirty. And after installing just five, the container design was scrapped in favor of traditional structures. Government officials apparently found the container design, so revolutionary for the MIT engineers, a symbol of poverty. Dominicans wouldn't be caught dead walking into one. "[T]he Lincos container was the brainchild of a group of Western and Western-trained technocrats," concluded researchers Paul Brand and Anke Schwittay in 2006, "They did not include indigenous designs, materials or needs into their broader design methodology, and the product of this methodology was ultimately rejected by the constituents the designers were supposed to serve."[9]

A Computer for the Rest of Us

Undeterred by Lincos's failure, in 2005 MIT Media Lab cofounder Nicholas Negroponte announced an ambitious project, One Laptop Per Child, with a bold goal: to deploy millions of laptops to children in the developing world, for less than $100 per unit. By 2012 the group had shipped some 2.5 million computers to more than forty countries.[10] Despite many setbacks, the project was considered a success by many, having spurred the development of a whole new class of low-cost laptops—netbooks.

Yet in the same time span, Nokia and its competitors sold over 2.5 *billion* mobile phones, nearly doubling the number of mobile subscribers worldwide from just over 3 billion in 2006 to 5.9 billion in 2011.[11] The transformation of the world's poorest continent is astounding. In Uganda, for instance, there are now more mobile phones than lightbulbs.[12] "Half of Africa's one billion population has a mobile phone," declared a 2011 headline in London's Sunday newspaper *The Observer*, "and not just for talking."[13] And in 2012, the rich world finally delivered an affordable computer to the developing world, when a price war in Kenya between South Korea's Samsung and China's Huawei drove smartphone prices there under $100.[14] One industry analyst believes that half the population of Africa will own one by 2017.[15] All across the globe, smartphones, rather than cheap laptops, are destined to be the true face of ubiquitous computing.

The economic impact of mobile phones has been transformative for the world's urban poor. A 2009 World Bank study of 120 countries found that for every ten percentage points increase in the penetration of mobile phones, GDP increased by 0.8 percent. The bank's chief economist, Christine Zhen-Wei Qiang, argues that "Mobile phones have made a bigger difference to the lives of more people, more quickly, than any previous technology. They have spread the fastest and have become the single most transformative tool for development."[16] For Nancy Odendaal, an urban planner who studies technology use in the townships of South Africa, "enabling livelihoods is the killer app" for these humble devices.[17] They have become indispensable tools for work, education, and health.

Developing countries have long struggled to build ubiquitous wired networks. In many places, as soon as telephone lines were laid, they would be torn out by thieves and sold as scrap copper. But wireless networks can be built faster and securely, allowing the benefits of connectivity to be quickly brought to large numbers of people. While the cost of building fiber-optic networks is thousands of dollars per home, delivering broadband wirelessly can cost one-fiftieth that much.[18] As a result, 80 percent of the world's mobile broadband subscribers are in developing countries.[19] Wireless is the infrastructure of inclusion—nothing else approaches the speed and cost with which we can now blanket entire cities with low-cost connectivity.

With the basic infrastructure of smartphones and mobile broadband in place, there has been an explosion in services aimed at the poor. Several innovation hot spots have emerged where start-ups are translating business ideas born on the desktop Web of the rich world into SMS-based services for megacities' poor.

In India, where one in six of the world's slum dwellers lives, mobile phones are creating tangible opportunities for work and education. Bangalore-based Babajob, in India's Silicon Valley, is an SMS-based social network for the millions of people working in the country's informal sector—day laborers, maids, drivers, and so on. One tech blog described the service as "LinkedIn for villages."[20] Another Bangalore nonprofit, Mapunity, emulates Google's sophisticated mapping services using people's mobile devices to sense traffic speed through phone movements and taxi radios. It then returns real-time traffic alerts via SMS.[21] South Africa's Dr. Math provides a tutoring service via SMS. Its American equivalent, the Khan Academy, requires an expensive laptop and high-speed Internet connection to access its recorded video lectures and chat rooms.[22]

In Kenya mobiles are the backbone of a new branchless banking system that is bringing financial services to millions for the first time. M-Pesa, named after the Swahili word for money, launched in 2007 and is now used by over 15 million people. Instead of building out a costly network of branches, or even automated teller machines, M-Pesa uses small retailers as its tellers. Through a secure process that confirms the electronic transfer in seconds, customers can withdraw or deposit cash with a few clicks. But as more of the country moves to electronic transfers, many transactions never even materialize as cash, flowing through the system entirely electronically. Safaricom, the country's dominant wireless carrier, created M-Pesa as a public-service initiative with a million-pound grant from the British government, and never expected it to turn a profit. Instead, it broke even in just two years and now delivers nearly one-sixth of the firm's revenues. During peak use, over two hundred transactions per second and 20 percent of Kenya's GDP streams through the M-Pesa network.[23] It is being rolled out across India, where it could eventually bring banking to hundreds of millions of poor people.

Most of the world's cities are now lit up by some kind of wireless service. But as Ericsson, a leading supplier of network equipment, points out, "Reaching the next billion subscribers means expanding to rural off-grid areas."[24] The company has developed highly efficient solar-powered cell towers for use in outlying areas where there is no electric-power infrastructure. On the consumer side, in 2010 Vodafone launched a $32 solar-powered phone in India.[25] Presumably, the arrival of modern telecommunications in the countryside might provide new local economic opportunities and slow migration to cities. But it could just as likely accelerate migration by plugging ever-larger rural areas into the social and economic life of the city. One study that tracked migration through mobile phones in Kenya uncovered an astonishingly high turnover rate for new arrivals—on average, newcomers during a year-long study period in 2008–2009 stayed in Kibera, the capital's largest slum, just less than two months.[26] Anthropologist Mirjam de Bruijn has documented Bedouin caravans in the southern Sahara that have altered their historic trade routes to periodically pass through areas of mobile phone service.[27] Even indigenous peoples want to stay connected in a global economy.

Development organizations are just beginning to wrap their thinking around the tremendous opportunity for development that mobile phones present. Richard Heeks, the professor of development informatics, sees a marked shift in the ICT4D movement from PCs to mobile devices. "We stand at a fork in the Internet access road," he wrote at the conclusion of his 2008 article. "We can keep pushing down the PC-based route when less than 0.5 percent of African villages so far have a link this way. Or we can jump ship to a technology that has already reached many poor communities."[28] It isn't just scholars and activists calling for a new model. In January 2013, when Google chairman Eric Schmidt spent a week visiting a handful of booming African cities, he saw firsthand the role of technology as a tool for economic opportunity. "This new generation expects more, and will use mobile computing to get it," he reported.[29]

Over the next decade, mobiles promise to become even cheaper and more pervasive. Assuming even a modest rate of replacement and a continued drop in smartphone prices, it is very likely that a decade from now half of the world's people—including hundreds of millions of the urban poor—will be walking around with devices that are essentially supercomputers in their pockets. Broadband wireless networks with data speeds in excess of 100 megabits per second or more will light up entire cities, including their slums.

But mobiles aren't simply new economic tools for the world's urban poor. Increasingly, mobile networks themselves are becoming observatories where we can watch in real time how people move, how cities grow, the quality of life, and economic activity.

Taking the Global Pulse

The lights went down on a room full of diplomats at the United Nation's General Assembly in New York in November 2011. "Imagine it's 2009, the rains are late, and food and fuel prices are rising," said Robert Kirkpatrick, director of the organization's Global Pulse project. "What would it have looked like in data collected by a mobile operator?"[30] He rattled off a list of telltale signs of distress. People might shift to smaller, more frequent purchases of airtime as their economic anxiety increased. Increased defaults on microloans would show up in payment systems like M-Pesa. Calls to livestock dealers would spike as families liquidated agricultural assets to survive. Phones purchased in villages would suddenly request connections with urban cell towers, as displaced farmers flooded the city looking for work.

The financial crisis of 2008 hit the world's poor hard. Food and fuel prices were already rising just as the contagion spreading through global financial markets released a parallel shock wave at the bottom of the pyramid. As Secretary-General Ban Ki-moon explained at the same event, UN officials were certain that the crisis would "inflict suffering immediately on the poorest and most vulnerable" people. But the economic chain reaction moved faster than his statisticians could track it. "It was clear we were seeing something new," he continued, "Impacts of the crisis were flowing across borders at alarming velocity." A decade's worth of economic gains evaporated overnight as hundreds of millions of families slipped back into poverty.

Ban Ki-moon moved quickly and decisively (by UN standards). "Our need for policy agility has never been greater," he explained, "Our traditional twentieth-century tools for tracking international development can't keep up. By the time we can measure what's happening at the household level, the harm has already been done." With the governments of the UK and Sweden serving as angel investors, Global Pulse was launched in April 2009 as the Global Impact Vulnerability Alert System (it was later re-branded). It was charged with developing new sources of real-time data to create an early warning system for social and economic crises.

Global Pulse promised to be the biggest advance in public demography in a generation. Consider the venerable US census, which is far more thorough and consistent than what poor countries can do. It swallows up huge amounts of resources and routinely misses millions of people. Since it is done only once a decade, there have only been a few dozen chances to improve it. While other less comprehensive interim surveys are taken to update the results, the master count happens just once per decade because it requires an army of over six hundred thousand census takers to collect data house by house. The United Nation's own methods are similarly plodding. As Global Pulse's 2011 annual report observed, "Traditional data collection methods like door-to-door household surveys . . . can take months or even years to complete and are woefully inadequate for this task."[31]

At the opposite end of the spectrum are the tools used by market researchers and pollsters. Free from the constraints that hamper government data collection, they can collect information almost anytime using any survey and statistical method at hand. They can tweak survey questions day-to-day to home in on emerging trends and fine-tune their observations. And they can go beyond surveys and tap nearly limitless pools of real-time private data on credit-card transactions, store visits, or web-browsing habits. Instead of sifting through the tailings of macroeconomic statistics for clues about recent events, they can plug into a sensory infrastructure that shows what is happening in the real economy at a microscopic level, second by second.

To bring the United Nation's crisis-sensing abilities up to date, Kirkpatrick partnered with a variety of research partners around the world to explore new ways of picking up signs of distress in the social and economic data exhaust of poor nations. One of the most promising experiments was done with Jana, a Boston-based company that had developed a tool for conducting surveys by mobile phone. Jana was the brainchild of MIT Media Lab alum Nathan Eagle, who spent several years in Kenya teaching students how to develop mobile phone apps. While working on a tool for nurses to report on blood supplies at rural clinics by text message, he noticed that participation quickly fell off. He needed a way to reward the nurses for responding to the text messages asking for updates on blood inventory. After returning to the United States, Eagle developed a system for compensating survey respondents with tiny amounts of airtime. Jana now has partnerships with hundreds of mobile phone companies, and can reach over 2 billion potential respondents worldwide.[32]

Global Pulse put Jana's system to work by sending out short queries by SMS, such as "Were you sick in the past 7 days?" or "If you had 15 USD what would you spend it on?"[33] The thousands of respondents were rewarded with free airtime from Jana's servers, which are wired directly into carriers' billing systems. Kirkpatrick insisted that these guerrilla surveys wouldn't replace traditional data collection efforts, but were intended rather to plug gaps and help inform the design of more traditional surveys. But if this approach proves accurate and reliable enough for day-to-day use, the data-gathering capabilities of poor nations could quickly leapfrog that of rich ones.

In another project, Global Pulse mined the web for real-time microeconomic signals. Working with PriceStats, a company that monitors online prices for some 5 million goods worldwide, researchers tracked daily prices for staples like bread, as opposed to the usual monthly government surveys. Surprisingly, this method even works in countries that lack widespread e-commerce. Even in countries with few Internet users, prices can still usually be harvested from online advertisements.

As promising as these new early warning networks were, Kirkpatrick was quick to temper the General Assembly delegates' expectations. "This is only a first exploration to confirm the potential of real-time data," he cautioned them. "We have not found the pulse, but we have a pretty good idea where to put our finger." To make the data actionable, Global Pulse created a collaborative website called HunchWorks that allows researchers, UN staff, and government officials to share insights about the data. Groups can create hypotheses using the data, score and discuss them, and then package up a dossier and mail the evidence off to a government and hopefully spur action.

Global Pulse is leading the development community's push into the next phase of ICT4D. Rather than pushing new technology onto the poor, it practices a kind of sensory jujitsu, leveraging the technology they are already using to better understand them. But in the end, that may limit its efficacy, for Global Pulse can only work in countries that invite the team in. Sadly, its ability to capture the plight of the poor in real-time, fine-grained details will be a hard sell in many nations whose governments have no wish to draw attention to their own failure to protect the poor and vulnerable.

Teach A Man to Fish . . .

Once upon a time, pedestrians in American and European cities lived in fear of airborne feces: before modern sanitation was introduced, the cry of "Gardez l'eau" (literally "Look out for the water!") would herald the evacuation of one's chamber pot into the street.[34] As cities like London boomed during the nineteenth century, every available body of water, from creeks to rivers to ponds, became an open cesspool. Only repeated cholera epidemics, and the "Great Stink" of 1858 (which forced Parliament to soak the curtains of the House of Commons in lime to mask the foul odor of the Thames River) would spur government action.[35]

Today, this ugly practice has reemerged for a whole new generation of city dwellers in the developing world, an ad hoc adaptation to unplanned urban growth and a lack of investment in sanitation. In the Kenyan capital of Nairobi, in place of chamber pots the residents of the massive Kibera slum have put the ubiquitous plastic bag to work. The process is much the same, however. Squat, step to the window, and hurl. Throughout the night, "Scud missiles," as the locals mockingly dub the flying waste packets, rain down on tin rooftops and hapless pedestrians. Compared to nineteenth-century London, the results are actually quite good. Sealed in their plastic tomb, disease-carrying microbes have a much harder time spreading. Cholera, dispersed through London's contaminated water supply, killed more than ten thousand people in 1853–54 alone.[36] Kibera has its share of water-borne disease but nothing on that scale.

Home to an estimated 250,000 residents, Kibera is one of Africa's largest slums.[37] But if you looked it up on Google Maps in 2008 and toggled between the satellite view and street-map view, you could make it disappear. One second it was there, a zoomable patina of corrugated tin shacks amid a rich tapestry of alleys and roads, unable to hide from a camera floating in space. Then it was gone, replaced by a blank spot drawn from a government map that still identified the area as the forest that previously stood there. Kibera's omission spoke volumes about how officials and the public saw it. Instead of the reality of a quarter-million people striving to build a future with their bare hands, all they knew were the sensational horror stories of Scud missiles.

Slums are often simply invisible to outsiders who lack basic information about who lives there. While Rio de Janeiro's top-down surveillance raises troubling questions about remote sensing of poor communities, the fact is that slums have much to gain from being documented. Being counted is the most basic act of inclusion—for a slum to assert its rights within the official city surrounding it, it needs to be measured and mapped. Many slum dwellers are taking matters into their own hands and arming themselves with new tools and methods to survey their own communities. Computerized mapping of cities is a half-century-old idea, originally developed by the US military and the census, but the first large-scale efforts to map slums didn't begin until 1994, in the Indian city of Pune. Led by Shelter Associates, an NGO formed by local architects and planners, "the project was based on the philosophy that poor people are the best people to find solutions to their housing problems," its founders wrote in the journal *Environment & Urbanization*.[38] Teaming up with Baandhani, an informal network of women who pool their savings to invest in better housing, the group surveyed slum residents, their homes, and the availability of fuel and electricity. In 2000 the city began funding the effort and in just two years it had surveyed some two-thirds of Pune's 450 slum settlements, mapping some 130,000 households.

The effort to put Kibera on the map was started by two geeks from the rich world, Erica Hagen and Mikel Maron, who in 2009 joined forces with a trio of Kenyan community-development groups to launch Map Kibera. They recruited a handful of twentysomethings who were active in the community, one from each of the slum's thirteen villages. With just two days of training in how to use consumer-grade GPS receivers, these volunteer mappers were sent out to traverse Kibera on foot, using their bodies as tools to collect traces of the thousands of streets, alleys, and paths that would form the first-ever digital base map of the thriving community. Results came quickly. "We did the first map in three weeks," Maron recalls.[39]

The mapping technique used in Kibera was imported from an unlikely place, which was also the source of the first modern surveys of Kenya—the country's former colonial ruler, the United Kingdom. Its development was the result of a dispute between the government and citizens. In the United States (and a handful of other countries, including Denmark and New Zealand), governments allow anyone to use a free digital version of their master street maps. But in the United Kingdom this data was tightly controlled by the government's cartographic agency, the Royal Ordnance Survey, which charged a fee for users until 2010.[40] This policy was widely regarded as a barrier to innovation, as it imposed substantial costs for amateurs, students, and others of little means who wanted to build new digital services that used maps.

By the early 2000s, artists and hobbyists in England had discovered that by plotting the position logs recorded by personal GPS navigators they could quickly collect the data required to re-create a digital base map of the street grid. Inspired by Wikipedia's model of collaborative knowledge production, in 2004 British computer scientist Steve Coast launched OpenStreetMap. Suddenly, anyone could upload a record of his or her movements along the nation's road network. By systematically traveling the streets of every city, town, and village in the United Kingdom, an army of volunteers set out to make a freely-usable map. As of 2013, after years of collective surveying and annotation, the crowdsourced street map of England was finally nearing completion. The effort has since expanded around the world, and in poor countries often rivals the government's own maps. After the 2010 Haiti earthquake, which obliterated the nation's mapping agency in a building collapse, OpenStreetMap provided essential data to relief organizations.

The Indian activists who pioneered slum mapping in the 1990s saw their work as a way to begin integrating poor communities into existing city-planning efforts in the hope of securing a fairer share of government resources. But with the new chart living online in OpenStreetMap, Map Kibera is focused instead on powering new tools that change how the community is represented in the media, and how organizers lobby the government to address local problems. Voice of Kibera, for instance, is a citizen-reporting site built using another open-source tool called Ushahidi. The name means "testimony" in Swahili, and it was developed in 2008 to monitor election violence in Kenya. Voice of Kibera plots media stories about the community onto the open digital map, and allows residents to send in their own reports by SMS. Another Map Kibera effort recruits residents to monitor the progress of infrastructure projects. Government-funded slum upgrades, such as the installation of water pumps and latrines, are hot spots for graft in Kenya. Many of the projects are awarded to friends of parliament members, and the government doesn't effectively monitor or audit contractors. Using this tool, residents can post reports on the actual state of construction, frequently contradicting the government's own claims. Over time, slowly but surely, the map is helping shift public perception of Kibera away from flying bags of crap and toward a view of a community of real people. As Maron told me, "People like living in Kibera. What they don't like is having raw sewage running by their house."[41]

Map Kibera represents a shift in how we think about using technology to help poor communities. We can ship all of the laptops we want to the world's slums, but we can't force anyone to use them, and even if they do we certainly can't guarantee it will have the intended impact. The United Nations can track all of the weak signals of economic distress from afar through efforts like Global Pulse, but the tools to intervene once a crisis is identified haven't changed much from yesteryear. Map Kibera demonstrates how open-source tools, put in place on behalf of poor communities, can empower them to create knowledge relevant to the problems they face. As Hagen described Map Kibera in a 2010 article, "It is founded on the premise that the advent of the digital age means that gatekeepers to information and data can often be bypassed or ignored completely, allowing for a new and sometimes parallel information system to be created and used by marginalized citizens."[42]

Since the 1990s, ICT4D projects mostly operated on an approach that Richard Heeks calls "pro-poor." As he puts it, in projects like One Laptop Per Child, "innovation occurs outside poor communities, but on their behalf." Truly sustainable solutions require people to participate in a project's design and implementation. Heeks calls this model "para-poor": outsiders work alongside members of poor communities in "participative, user-engaged design processes."[43] As the movement evolves, and technologies like the mobile phone trickle down, Heeks envisions a second shift to "per-poor" innovation, done entirely by and for the poor. While Map Kibera is clearly a para-poor project, with Westerners bringing in new technology and design ideas, it has created a framework on which per-poor innovation can happen.

Mapping has tremendous power to improve the slums of the developing world. John Snow's map of cholera deaths in 1850s London recast the public understanding of slum conditions, and spurred reforms that eventually rid the city of the disease for good. In India, slum mapping is helping change the practice of city planning, which long considered those communities "chaotic masses rather than coherent urban areas," according to Shelter Associates.[44] But in both cases, governments responded achingly slowly. Map Kibera offers the hope that by using maps to power community-based initiatives, rather than simply lobby government, progress will be faster.

It's unconscionable that governments continue to ignore the slums, to pretend that they are invisible. But in sub-Saharan Africa, where the UN estimates that six out of ten people live in slums, with a little help from concerned outsiders slum dwellers are rewriting the map themselves.[45] But it won't happen on its own. Map Kibera's lesson is clear—it isn't enough to simply drop technology into poor communities. Do-gooders will have to stick around long enough to teach people how to use it. "Give a man a fish and you feed him for a day," the Chinese proverb goes, "teach a man to fish and you feed him for a lifetime."

From Digital Divide to Digital Dilemmas

While the simple rubric of the "digital divide" has been used for nearly twenty years now to frame policy debates about technology and the poor, it is no longer useful. The problem isn't just access to technology; it is the lack of capacity to exploit it for good. As the World Bank argues, "Not all economies are the same and not all economies are equally prepared to absorb broadband and embrace it to reap its potential benefits."[46] Thinking simply of a digital divide tricks us into believing this is a simple binary problem of haves and have-nots, when in fact it is a set of interlocking dilemmas that defy easy solution. This is as true in the poor parts of the developed world as it is in the developing countries. Moldova and Detroit, Kibera and Cleveland share many similar challenges in realizing the potential of smart technology.

The first dilemma concerns access and agency. Putting technology in the hands of the poor, as OLPC did, is one step. But expecting that access alone will create opportunity is no longer appropriate. Helping the poor secure the skills and support to make use of it is far more challenging. This was one of the most painful lessons from the first generation of ICT4D efforts, and the problem is endemic to smart-city projects, not just those in the developing world.

Take 311 telephone hotlines, which have become a widely used means to access government information and services. On the surface, they appear the most universally accessible of all smart-city systems, with few of the barriers that hinder use of Web-based tools or mobile apps: 311 services run over the nearly ubiquitous telephone network, are open twenty-four hours a day, and are typically offered in many languages. In New York City, whose 311 system averages some sixty thousand calls each day, more than 170 languages are offered.[47] A resident can use 311 to interact with government without even knowing how to read or write. It would be difficult to design a more accessible system. But 311 has its own secret digital divide. According to a 2007 study conducted by Columbia University for the New York City Department of Sanitation, poor neighborhoods with large minority populations complained less frequently to 311 about missed trash pickups.[48] And New York isn't alone in its underutilization of 311 by historically disadvantaged groups. When I visited Vancouver in 2011, City Councilor Andrea Reims explained to me how that polyglot city has had a similar experience with its large Cantonese-speaking population. The reasons why non-native English speakers do not use the system are not well understood, but presumably they stem from unfamiliarity with this new way of interacting with government, legitimate and/or irrational fears among immigrant communities about government, or different cultural norms for how issues are dealt with at a local level. Nonetheless, the result is the same. Native English speakers are complaining more, and their complaints are being used to disproportionately dispatch resources to address their problems. Layered on the injustice is the fact that native

English speakers are already more well off: they tend to be better educated and have higher incomes.

The point is that great vigilance is needed to ensure that smart systems don't create new exclusions. Development economists used to measure poverty solely on the basis of per capita income. Today, they increasingly use multidimensional measures that paint a richer picture of health, education, and living standards.[49] To truly understand what prevents poor people from making use of technology we will need to develop multidimensional assessments of technology and information literacy.

Another dilemma will revolve around the use of big data in real time, as systems like Global Pulse start to inform decisions in everything from urban planning to aid programs to disaster relief. It is one thing for data to render a problem visible, and quite another for that data to inform a response. Instead of reducing the role of guesswork and intuition, big data might create even greater uncertainty. In everyday situations, leaders who don't understand or trust the data will simply fall back on their instincts. Worse, during a crisis, the pressure to act decisively could lead to inadvertent use of immature data and a rush to improper conclusions.

Deep, large-scale sensing of data about populations creates its own dilemma—the need to balance the privacy rights of individuals and small groups against the larger public good. Every society will have to find its own balance. While Kibera highlights the risks of being left off the official map, in many cases the poor may resist external efforts to measure and manage their communities. Global Pulse takes great pains to explain its data-privacy precautions, presumably knowing full well that in many countries its data-gathering tools could rival those of national intelligence agencies. New techniques of monitoring populations in order to help them could be copied or co-opted by governments looking to subdue them.

The most gut-wrenching question is whether haves should play a role in changing the fate of have-nots. For the past several decades, the goal of aid programs has been to modernize poor communities and bring them up to the standards of the rest of the world. Many of these efforts have failed, often because they didn't take into account existing knowledge and assets in poor communities. The spread of cheap smartphones, fast wireless networks, and open data—along with the skills to make use of these tools—will be a boon for self-propelled development. Slum dwellers are incredibly facile in upgrading and improving their homes and infrastructure with the most basic of resources. A parallel digital effort is likely to produce just as much innovation. But the independent bootstrapping of smart slums, however romantic or politically incorrect, seems unrealistic. And there will always be an urge to "do something," if only for self-preservation. As Heeks argues, "In a globalized world, the problems of the poor today can, tomorrow—through migration, terrorism, and disease epidemics—become the problems of those at the pyramid's top."[50]

This brings us to the final dilemma: crowdsourcing and the future role of government in delivering basic services. In smart cities, there will be many new crowdsourcing tools that, like OpenStreetMap, create opportunities for people to pool efforts and resources outside of government. Will governments respond by casting off their responsibilities? In rich countries, governments facing tough spending choices may simply withdraw services as citizen-driven alternatives expand, creating huge gaps in support for the poor. In the slums of the developing world's megacities, where those responsibilities were hardly acknowledged to begin with, crowdsourced alternatives may allow governments to free themselves from the obligation to equalize services in the future. As fashionable as it has become in the developed world, crowdsourcing is highly regressive. It presumes a surplus of volunteer time and energy. For the working poor, every second of every day is devoted to basic survival. The withdrawal of any government services would remove a critical base of support for these extremely vulnerable communities.

For engineers and technologists, the intractability of these dilemmas is deeply uncomfortable. Information technology has remarkable power to help the poor help themselves, but to date its greatest impact has been to lure them off their farms to squatter cities where they now wait to see if they'll be permitted to grow rich too. Whether or not they do, the democratization of smart technology is certain to allow poor communities to pursue their own vision of a smart city. Across Africa, more than fifty "tech hubs, labs, incubators and accelerators" have opened doors in recent years, according to the BBC. Nairobi has six alone.[51] Inevitably, what happens in these new centers of innovation will shape the way we think about the place of technology as well. For I have no doubt that right now, somewhere in Kibera or Soweto or Dharavi, some young civic hacker is cobbling together a few transformational bits of technology that will change the world.

7

Reinventing City Hall

Whether it's built by big companies or wireless activists, the first prerequisite for entry into the club of smart cities is a world-class broadband infrastructure. Over the last decade, a growing number of cities have tried to speed the process and introduce competition by building new networks themselves. But across the United States, the telecommunications industry has fought these civic initiatives to a standstill. Perhaps it is fitting that one of the first battles over the smart city took place in Philadelphia, the birthplace of American democracy

"Forget cheese steaks, cream cheese and brotherly love," the *New York Times* gushed. "Philadelphia wants to be known as the city of laptops."[1] On March 5, 2004, Mayor John Street stood before a crowd at Love Park in Center City to inaugurate the first hot spot of Wireless Philadelphia, an ambitious project to blanket the city's 135 square miles with low-cost Wi-Fi.[2] At the time, a handful of smaller cities—such as Long Beach, California—had built public wireless networks in their downtowns. But Philadelphia was the first major American city to aim for ubiquitous, citywide coverage. Street, himself a technophile, saw the network as an engine of rejuvenation for the economically depressed city. As Greg Goldman, the project's former CEO, reflected some years later, the whole point was "to make Philly a cooler place to live. John Street understood the power of technology and getting it into the hands of the neighborhoods."[3]

The city was soon abuzz. A 2005 *Philadelphia Magazine* cover story on the city's resurgence boasted, "the Street administration's plan to turn the entire city into a Wi-Fi hot spot of low-cost wireless Internet access has generated more positive attention than anything we've done here since 1776."[4] It was a bold plan, seemingly without political risk for Street or financial risk for the city. The network's projected cost was just $10 million, to be raised entirely from private sources. Work was to start within the year and be completed in just twelve months. For Goldman, it promised to mark a transformational moment for the city. "It had tremendous political support, it had private capital driving its expansion, it had a tremendous degree of public engagement, it even had strong media support," he says. The formula was rapidly copied by San Francisco, San Diego, Houston, Miami, and Chicago. As other cities followed Philadelphia's lead, it seemed an endorsement of the plan.

The city quickly inked a deal with Internet service provider EarthLink to push the public–private partnership forward. After growing into one of the largest dial-up purveyors during the 1990s, EarthLink was trying to get out of that rapidly declining business and elbow its way into the broadband market. Federal telecommunications reforms enacted in 1996 to increase competition had ordered regional telephone companies (the "Baby Bells") to provide competitors access to their new high-speed digital subscriber lines (DSL). But in practice the Bells were slow to process requests for access, creating long installation delays for companies like EarthLink, which struggled to gain market share. Tacking, the company placed a bold bet on municipal Wi-Fi as a way to deliver broadband directly to homes and businesses. The design for Philadelphia now called for a $20 million build-out, lighting up 80 percent of the city with over 3,500 light-pole-mounted transceivers, according to Goldman.

Philadelphia's love affair with Wi-Fi quickly turned sour. "Everything that the project had going for it, turned against it," Goldman laments. Street's administration was hamstrung by corruption scandals that led to increased criticism of all his initiatives. EarthLink, in a desperate bid for survival as its dial-up business collapsed, took on too many wireless projects in other cities. Work on the Philadelphia project slowed to a crawl. In early 2007 the company was thrown into disarray by the sudden death of its longtime leader and CEO, Garry Betty.

Deployment challenges dogged the project at every turn. In Center City, Philadelphia's gentrified urban core, historic-preservation regulations prohibited mounting the antennas on the district's ornamental streetlights. The hilly, tree-lined streets of the city's more affluent outlying neighborhoods were a wireless engineer's nightmare. These complications left unserved the two enclaves where the city's political power brokers lived. PECO Energy, a unit of the energy giant Exelon, owns Philadelphia's light poles and charged steep rates for hosting the wireless modules. In Goldman's words, the company "was very, very difficult. They took their money but they were not a partner."

But Wi-Fi technology's own limitations doomed Wireless Philadelphia in the end. As we saw in chapter 4, Wi-Fi was never designed for large-scale, seamless outdoor networks, let alone delivering broadband to building interiors. Goldman later confessed, "we bought EarthLink's promises hook, line and sinker." Around the United States, other communities were running into similar problems. Lompoc, California, a rural community of forty-two thousand, had installed citywide Wi-Fi as a revitalization strategy after cutbacks at a nearby military base. It soon discovered that wire mesh embedded in the town's numerous stucco-sided homes blocked wireless signals from reaching devices inside.[5] Goldman believes that Wi-Fi was the project's Achilles' heel in Philadelphia. "If the technology had worked fine, we would have been able to weather all of these other problems." By 2008, the *New York Times*'s excitement had cooled. "Hopes for Wireless Cities Fade as Internet Providers Pull Out," read the March 22, 2008 headline.[6]

While Philadelphia inspired many other cities to launch their own wireless initiatives, it also provoked a vicious response from telecommunications companies, setting back the prospects for municipal broadband throughout the United States. Horrified by the prospect of competing for customers with local governments, within months of Street's ribbon cutting, the industry unleashed its counterattack. Verizon, the dominant local telephone company in the city, lobbied the Pennsylvania legislature to pass a law barring cities from charging any fees to recover the costs of building municipal broadband networks. In a last-minute compromise, Philadelphia was grandfathered in and allowed to complete its project. But it would be the last municipality in Pennsylvania to build a public broadband network.

Industry lobbyists fanned out across the country, and with Pennsylvania's law as a template, succeeded in getting roadblocks passed in half of the nation's states.[7] At the time, Federal Trade Commissioner Jon Leibowitz, speaking at the National Association of Telecommunications Officers and Advisors conference, said, "imagine if Borders and Barnes & Noble, claiming it was killing their book sales, asked lawmakers to ban cities from building libraries. The legislators would laugh them out of the State House. Yet the same thing is happening right now with respect to Wi-Fi and other municipal broadband plans, and it is being taken all too seriously."[8]

The potent and effective industry backlash to Wireless Philadelphia means thousands of American communities are now restricted by state law from investing in their own future. And telecommunications companies continue to vigorously fight local efforts. In Colorado, a somewhat less restrictive bill passed in 2005 that permits municipalities to build broadband only after approval by a public referendum. In the city of Longmont, a cable industry–backed lobbying group called Look Before We Leap spent $300,000 on advertising to unsuccessfully stop a 2011 municipal referendum to fund a city-owned fiber-optic network.[9] As Vince Jordan, who runs the project for Longmont's electric power utility pointed out in a podcast soon after voters approved the measure, it was the most money ever spent on a local campaign in the 86,000-person city's history.[10]

Goldman believes that it is unfair to depict Wireless Philadelphia as an utter failure. "This was a beta project for all that cities are trying to do today," he says. Broadband prices in Philadelphia dropped immediately as soon as the project was announced, as they do everywhere that local governments enter the broadband market. Over two thousand low-income families received free laptops and discounted Internet service from the project's digital inclusion initiative. The battle for Philadelphia also eventually provoked cable giant Comcast, whose corporate headquarters is one of the city's largest employers, to roll out its own free Xfinity Wi-Fi service up and down the Eastern Seaboard. Since the company had a near-monopoly on cable television in its markets (the regional telephone giant Verizon's own TV offerings were only slowly making inroads), this free perk for Comcast subscribers became a kind of de facto public Wi-Fi network.

In the end, Philadelphia did get its wireless network after all. Unable to make the business work in cities across the United States, in May 2008 EarthLink announced it was getting out of municipal wireless entirely and liquidated its assets in Philadelphia. Less than two years later, the city belatedly decided to buy the network back from a bankrupt holding company that had acquired it in 2008, after EarthLink had unsuccessfully tried to give it to the risk-averse city for free. The final cost, a mere $2 million.[11] Philadelphia's wireless network will now be repurposed for public safety and government operations, linking up video surveillance cameras and city workers' handheld devices. After so much turmoil, it was a fire-sale bargain for the city. New York City, by comparison, spent a whopping $549 million building out its own public-safety wireless network, which costs more than $38 million a year just to operate.[12]

Luckily, Philadelphia was able to salvage value from the debacle. In 2007, when the projects in Philadelphia and Lompoc were falling apart, I told a reporter from the Associated Press that municipal wireless networks "are the monorails of this decade: the wrong technology, totally overpromised and completely undelivered."[13] That was true, but the lessons of Philadelphia have led to more successful efforts in dozens of other communities around the United States. Communities that pursue Wi-Fi today, like Chattanooga, Tennessee, are doing it in a more systematic, targeted, and understated manner—and it is often an add-on to a more robust fiber-optic network rather than a substitute for wired connectivity. Which is good. Otherwise, they're just legacy projects—like convention centers, casinos, and sports stadiums—for headline-grabbing mayors looking for easy wins.

Fishing for Apps

Chastened by the struggles to build municipal wireless networks and hampered by chronic budget shortfalls, many cities today are seeking less risky ways to experiment with smart technologies. In recent years a growing number have tapped software firms and freelance hackers in their own backyard, fishing for useful apps with government data and cash prizes as bait.

It all began in Washington, DC. The home to an increasingly dysfunctional national government, as a city Washington is still recovering from the calamitous reign of former mayor Marion Barry, who in the 1990s served six months in prison for crack possession between his third and fourth terms in office. In the last decade, however, the DC metro area has grown into one of the nation's most important high-tech hubs, second only to Silicon Valley in total tech employment. While most of the jobs are in the suburbs, lots of young software engineers now live in the district's gentrified neighborhoods around Dupont Circle and in Adams Morgan, working at start-ups, government agencies, and non-profits.

By 2008, the tide was turning in DC. In his first year in office, mayor Adrian Fenty had restructured the city's schools and expanded community policing, driving down crime. Technology played an important role, allowing informants to send text messages to the police department anonymously. After releasing hundreds of government data sets early that year on a new city website, the DC Data Catalog, Fenty's chief technology officer Vivek Kundra worked with local tech community organizer Peter Corbett to design a contest around the site. Apps for Democracy, which launched in October, challenged the local tech community to create software that would exploit this new public resource. The city put up a $50,000 purse to sweeten the pot.

In just thirty days, local citizen-programmers created forty-seven different Web and smartphone apps that tapped the DC Data Catalog. Winning entries ranged from Point About, an iPhone app for receiving real-time alerts on crime, building permits, and other essential city operations, to DC Historic Tours, a Google Maps mash-up for making customized tourist itineraries out of Wikipedia entries and Flickr photos of historic places. Fenty cast the contest as a masterstroke of fiscal leveraging, announcing that "on the scale of how governments have traditionally done things, this is not an expensive program."[14] It was also blazingly fast. Kundra's office estimated it would have taken over a year (and up to two) for the city to have bought the apps through normal procurement channels. In a clever bit of recession-proof public relations, Corbett and Kundra calculated that the apps represented $2 million worth of in-kind services—a more than 4,000 percent return on the city's $50,000 investment.[15] Based largely on the success of the Data Catalog and Apps for Democracy, Kundra was tapped for president-elect Barack Obama's transition team and was appointed as the federal government's first chief information officer just four months later.

Apps contests and open city data spread quickly after DC's initial success. The low-cost combination was an irresistible tool for mayors facing growing demand for interactive services from smartphone-toting citizens, and an economic recession that decimated their budgets. As stimulus funding ran out and fiscal austerity took hold, it was a model that could deliver innovation with nearly zero funding. The needed data was already mostly online in many cities, but scattered across a constellation of government websites. All a city had to do was assemble it in one place. Within a year, New York, San Francisco, and Portland, Oregon, all launched similar efforts, and DC held a second round of Apps for Democracy in 2009. Over the next several years the idea spread abroad as Edmonton, Canada (2010), Amsterdam (2011), and Dublin (2012) followed suit. Meanwhile, the World Bank was exporting the model to the developing world through its own Apps for Development contest held in 2010.

The success of apps contests comes from their ability to quickly assemble technical teams that can repackage government data in novel ways that are valuable to citizens and local businesses. Many of the submissions have been mundane, and some simply esoteric, but a handful have really stretched the idea of what smart cities could be. Consider, for example, "Trees Near You," an entry in New York City's first BigApps contest. Among the usual data exhaust of government that powered the contest—health inspection grades and noise complaints—lay an odd database, the Street Tree Census. To set a baseline for the city's ambitious PlaNYC sustainability initiative, which included a drive to plant one million new trees, in 2005 the Parks Department had asked 1,100 volunteers to count every single sidewalk tree across the five boroughs and record each one's vital characteristics. From anywhere in the city, the Trees Near You app allowed iPhone owners to browse the database's logs for nearby trees, learning about their species, age, and ecological benefits. No bureaucrat would have ever dreamed up (or justified spending money on) what tech entrepreneur Lane Becker called "a beautiful, almost meditative iPhone app."[16]

As good as they were for brainstorming and stretching the notion of the possible, apps contests have produced few scalable, sustainable successes over the long run. Of the hundreds of apps submitted in the first two BigApps contests in New York, just one received any significant venture-capital financing to continue its work—a clunky city guide called MyCityWay that was basically just a browser for many of the city's newly public data sets. (And it was so-called dumb money, $5 million from BMW's i Ventures arm, a newly launched strategic fund whose management lacks the deep industry knowledge and connections that entrepreneurs value highly in investors.) The winner from 2010, a crowdsourcing transit app called Roadify, has received some angel funding. Trees Near You developer Brett Camper moved on to other projects. The app sat there in the iTunes store, frozen, un-updated in its original state, until it was finally removed in late 2012. In fact, the vast majority of apps entered into contests are quickly abandoned. As John Geraci of DIYcity points out, city apps contests "are very good at producing version 1 apps, when what a city government needs is the rock-solid, full-featured version 7."[17]

The real problem with apps contests driven by new government data, as we have seen, is that they rely on programmers to define problems, instead of citizens or even government itself. The only requirements of that first wave of city apps contests, and a still-surprising share today, was that it use one of the data sets released by the city. But as New York–based interaction designer Hana Schank wrote in a scathing critique on the eve of New York's third BigApps contest in 2011, "website and app development begins with a deep look at what the end users need, and how they are likely to use sites and apps in the course of their day. The problem with the BigApps contest is that it leaves both user needs and likely user behavior out of the equation, instead beginning with an enormous data dump and asking developers to make something cool out of it."[18]

The data-centrism of city apps contests is all the more curious because it ignores the key incentives of the wildly successful philanthropic grand challenges that inspired them. The Ansari X PRIZE, the granddaddy of modern innovation contests, challenged competitors to build a reusable spacecraft that could fly twice in one week, an unheard-of feat. By defining a single difficult problem, it captured the imagination of the nation's brightest engineers and most ambitious entrepreneurs, leveraging $100 million in privately funded research with just $10 million in prize money. Less than eight years later, Burt Rutan's SpaceShipOne touched down in the Mojave Desert to win the purse. The money mattered, but the prestige and breadth of accomplishment were the real motivators.

Subsequent rounds of the apps contests in Washington (in 2009) and New York (in 2012) did add a problem-definition round, challenging a larger group of citizens to tell developers the kind of apps they wanted. But beyond crowdsourced voting, there was no process to winnow the pool of ideas down to a few truly important problems. Programmers were encouraged to troll the discussions for app ideas, but not required to address them at all. Not until the fourth annual BigApps contest in 2013 did New York City finally engage a variety of partner organizations with deeper knowledge of its citizens' most pressing problems to define briefs on what it called BigIssues in four categories—jobs, energy, education, and health.

In retrospect, Fenty's claim that the original Apps for Democracy campaign saved millions was also deeply misleading. None of the apps responded directly to a pressing civic need, which meant the city probably never would have spent the money to create them without the contest. And none of the apps contests to date have dictated that entrants hand their code over to government or even open-source it—in the end, it has been up to cash-strapped developers to maintain the software and any server infrastructure it requires.

Apps contests also highlight the gap between haves and have-nots in smart cities. In 2010, less than two years after Apps for Democracy launched, Washington's new chief technology officer Bryan Sivak scrapped the contest. His glum assessment: "If you look at the applications developed in both of the contests we ran, and actually in many of the contests being run in other states and localities, you get a lot of applications that are designed for smartphones . . . devices that aren't necessarily used by the large populations that might need to interact with these services on a regular basis."[19] *The Hill*, a popular DC political blog, scornfully reported, "The contest is just the latest of Kundra's efforts as D.C. CTO to come under greater scrutiny since his departure . . . none of his projects seem to have made a lasting impact on the District's government."[20]

It wasn't just the focus on smartphones that left regular people out, however. Without a formal process to connect programmers to a representative group of citizens, unsurprisingly the contests tended to produce apps that solved the problems of a connected elite. Moreover, the crucial challenge of rendering and promoting successful apps in multiple languages and ethnic communities has been utterly neglected in apps contests.

The one clear sweet spot for city apps has been public transit. All transit operators face the thorny problem of communicating schedules, delays, and arrival information to millions of riders. Apps provide a quick, cheap, flexible, intuitive, and convenient way to push both schedules and real-time updates to anyone with a smartphone. As of early 2012, over two hundred transit agencies in North America were publishing some form of schedule information using a machine-readable format called General Transit Feed Specification, developed in 2005 by Google engineer Chris Harrelson and Bibiana McHugh, a technology manager at Portland, Oregon's Tri-Met transit authority.[21]

Unlike most contest-generated apps, transit apps have a huge preexisting market, making it possible to build viable businesses that leverage open government data. Francisca Rojas is a researcher at Harvard University's Kennedy School of Government who has studied the impacts of open transit data. As she explained it to me, "The difference with transit data is that developers are maintaining and improving the apps rather than abandoning them. Users are willing to pay for transit apps and continually suggest new features to developers to make them better, and transit agencies keep releasing new and improved data sets."[22]

Investing in transit apps is also good public policy. They're highly inclusive and the benefits accrue to the working poor who depend on public transportation the most. For a working mom struggling to balance childcare and a long commute, knowing the arrival time of the next bus is a huge help. And as apps make transit easier to use, they might help tempt drivers out of their cars and onto buses and trains, where they can be distracted by their online lives more safely and productively even as they cut their carbon emissions.

Cities are also moving to create apps to address specific problems. For instance, the hilly city of Bristol, England, commissioned Hills Are Evil!, an app that "provides people with restricted mobility, cyclists, skateboarders, the elderly, and people pushing pushchairs, the ability to identify the most appropriate route between two places."[23] As a result of its experience with apps contests, in 2011 New York City's internal technology department began to explore reforming how it competitively bids small software projects to allow it to more rapidly source apps from small businesses and individuals.[24]

Ultimately, apps contests are having a positive long-term economic impact, regardless of whether they deliver useful technology. They have catalyzed a community of technologists inside and outside government who are committed to improving the lives of residents and visitors. Instead of working at cross-purposes, or viewing each other with distrust, hackers and clued-in bureaucrats are learning how to work together to prototype new approaches to old urban problems and explore strange and exciting new possibilities.

Data Junkies

Mayor Rudolph "Rudy" Giuliani tamed New York City, a metropolis once thought all but ungovernable, through the blunt force of law. His successor, Michael Bloomberg, whose business empire was built on the delivery of financial data to traders around the world, was a technocrat who rules through scientific management. "If you can't measure it, you can't manage it," he was known to say.

In the spring of 2010, soon after beginning his third term in office, Bloomberg enlisted the help of Stephen Goldsmith to ensure this bean-counting legacy. The former mayor of Indianapolis, Goldsmith took over as deputy mayor for operations, a position with broad authority over the city's police, fire, sanitation, and buildings departments. He arrived with a reputation for privatizing public services and busting municipal unions. In Indianapolis during the 1990s he had reduced the city's payroll by nearly a quarter by letting private companies compete against city departments for dozens of services such as repairing potholes and washing fire engines.[25] John Hickenlooper, the two-term mayor of Denver, Colorado (2003–11), put it best when he said, "The most important thing a mayor does is hire talented people to run the city." Goldsmith was a hired gun brought in by Bloomberg to simplify and streamline government.

In June, just two months after Goldsmith arrived in New York, I listened as he laid out his vision in a brainstorming session with local techies and e-government wonks held at the Harlem headquarters of Living Cities, a club of foundations active in urban issues. In Goldsmith's view, a century's accretion of rigid procedures, inflexible work rules, and mindless checklists were preventing city workers from developing critical thinking skills and the ability to make decisions in the field in response to citizens' needs. Taking the city's Department of Buildings as an example, he explained how data mining could empower them to think on their feet, and react rapidly to changing uncertainties instead of mindlessly ticking off boxes on a checklist. Starting with an analysis of risk factors, a piece of software could prioritize each day's inspections instead of just working through a sequential list of addresses on some rigid calendar. Then, during the actual inspection, another analysis would point out the most likely trouble spots that needed scrutiny by the inspector's expert eye. Goldsmith wanted to turn city workers from automatons into knowledge workers.

The stated goal of this approach was an increase in productivity and effectiveness. But as with his privatization efforts in Indianapolis, it was also a Trojan horse for an assault on the city's powerful labor unions. Fully implemented, Goldsmith's reforms would make redundant an entire swath of middle managers, the supervisors and dispatchers who jockeyed line workers through their daily procedural paces.

This Hoosier was in for a New York City street fight. A few months later Goldsmith announced a streamlining plan for the Department of Sanitation, cutting four hundred jobs through attrition and demoting a hundred supervisors back to the line. The timing could not have been worse. Over Christmas, a blizzard struck New York City. Goldsmith was out of town and waffled on declaring a snow emergency. Although accusations of wildcat strikes and work slowdowns by sanitation workers who manned the plows were never substantiated, parts of Queens remained unplowed for days. It was a stunning replay of the infamously botched cleanup after the 1969 blizzard, an event many chalked up as union retaliation for Mayor John Lindsay's rough tactics during an earlier strike.[26] Goldsmith never recovered politically from the debacle, and resigned the next summer after just fourteen months in office.[27]

Snow also set the stage for another data-driven mayor to take office in Chicago. Just one month after New York's blizzard, Chicago mayor Richard Daley faced down an even worse snowstorm. In another bungled response, Daley's chief of staff delayed a decision to close down Lake Shore Drive, and hundreds of people were stranded as cars and buses were trapped by drifting snow. Daley faced some of the harshest criticism of his twenty-two-year reign.[28]

When Rahm Emanuel, former White House chief of staff and mayor-elect, arrived at Chicago's City Hall in May 2011, the memory of the fiasco was still fresh. As summer turned to fall, forecasters predicted a harsh winter ahead. Unlike New York, where plowing progress was tracked manually by radio reports from drivers during the 2010 blizzard, Chicago had installed GPS trackers on all its plows in 2001.[29] City officials could follow the plows on a real-time map, but citizens had no way to access this information. Accusations of preferential snow removal on streets and in neighborhoods of the mayor's political supporters were common.

The lack of transparency around Chicago's plowing operations is far more typical of how city governments operate than the free-for-all data giveaways of apps contests. The vast majority of the data that city governments collect remains hidden. Department heads guard this data closely, and resist sharing it even with each other, let alone with the public. It is the source of their power, and it can expose their shortcomings.

But as Emanuel said in the weeks following Barack Obama's election in November 2008 amid the global economic meltdown, "You never want a serious crisis to go to waste. . . . This crisis provides the opportunity for us to do things that you could not do before."[30] John Tolva, Emanuel's new chief technology officer, had a simple solution—open up the plow map. The result, Chicago Shovels, sported a gamelike Plow Tracker map that showed the progress of plows during major storms. But Tolva also saw the map as a way to recruit citizens to help with snow removal and developed a tool called Snow Corps to match shovel-ready volunteers with snowbound senior citizens. Tolva's approach to data-driven reforms couldn't have been more different than Goldsmith's in New York. Instead of data-mining organizational charts and performance to right-size the city workforce, he opened up operational data to mobilize citizens. And he had his own ideas about using technology to make government more cost-effective.

Tolva's path to public service began on the windswept platform of the L, as the city's elevated trains are called. As he recalls, "During the mayoral campaign, Rahm did a tour of over a hundred L stops. It was December, it was freezing, it was early and I went into my L stop. He and I were the only people there, so I approached him and said, 'What do you think about open data?'" Emanuel countered, "Do you mean like, transparency?," Tolva told me.[31]

"If I was going to hook him I would have to hit him where it hurts," Tolva recounts. "No, I mean saving money." People streamed into the station around them, but Emanuel ignored them, momentarily fixed on Tolva. Before turning to greet the throng of prospective voters flooding into the station, he locked hands with Tolva. "We should talk," the candidate told him. Five months later, Tolva received an invitation to join the mayor-elect's transition team.

Bloomberg may be fond of numbers, but Tolva is a data junkie, obsessed with the stuff and always on the hunt for more. Before taking the job as Chicago's chief technology officer, he had spent some thirteen years at IBM—most recently as the head of the company's City Forward project, an effort to evangelize the virtues of data-driven decision making in local government. One of the projects he oversaw was the deployment of City Forward's Web app, which let people create benchmark comparisons between cities around the world using a variety of vital statistics.

By early 2012 Tolva was working hard to live up to the promise he'd made to the mayor on that train platform. He was busy building an early warning system of his own, like the UN's Global Pulse, to scour the city's data for trouble spots. As we spoke by phone, he overflowed with excitement about all of the free technology at his disposal, rattling off a laundry list of powerful open-source software tools that were rapidly democratizing the ability to manage and analyze big data. They include MongoDB, a tool for managing huge databases (which Tolva learned about from the Foursquare crew) and R, a language for statistical analysis.

Early results of these number-crunching explorations of the city's big data are tantalizing. "Deep analytics," he says, borrowing IBM's jargon for the collection of tools and techniques for dissecting big data, "is about more than more than just performance management and transparency. It's about showing us where there are connections that we did not realize." In one experiment, his team cross-referenced Meals on Wheels delivery logs with the city's own tax records to generate a map of elderly living alone. "We can start to build up a list of people that need to be checked on during heat and cold emergencies," he says; "Is that a cost saving tool? Yes. But it is also a lifesaving tool."[32] In Chicago's harsh climate, extreme weather routinely claims the lives of dozens of seniors.

Inspired by popular data-driven online indexes like WalkScore, which computes a numerical measure of walkability for any US street address, Tolva was also working on a Neighborhood Health Index. A massive mash-up, it would synthesize "all the indicators that we have block by block and infer the probability that an undesirable outcome will result." While Chicago's effort looked at real data, not some abstract model, there was an eerie similarity to the cybernetic missteps of the 1960s that tried to compute urban decay. But Tolva wasn't entirely seduced by data. He understood that it is nothing more than a diagnostic tool: "A single data point that does not tell you that a house is going to fall into blight but [the index could signal] that there is a higher than normal probability that it will be in disrepair."[33] The data could then be used as an input when allocating revitalization funds or directing social workers to trouble spots. It was a strategy cut from the same cloth as Goldsmith's vision for transforming bureaucrats and civil servants into knowledge workers, but without the union busting.

As a triage tool for stretching scarce city resources, it's hard to argue against this kind of data-driven management. But as data becomes more central to how we measure government performance, it can create perverse incentives. One of the largest and longest-running data-driven management systems of any American city is the New York City Police Department's CompStat program. Since 1994 CompStat has combined computerized mapping of crime reports with weekly roll-call meetings where commanders are grilled by their superiors over any errant localized spikes in lawlessness. In practice, it allows the NYPD to shift resources to wipe out crime hot spots before they can undermine a community's sense of order. For many years, the program was widely credited for the stunning decline in New York's crime rate in the 1990s, though many other theories have been put forth to explain it (for instance, the reduction in the number of at-risk teens following the legalization of abortion decades earlier, and the end of the crack epidemic). Regardless of its efficacy, in recent years criticisms of CompStat's impacts on policing have mounted.[34] It turned out that, in their quest to maintain steady reductions in the reported rate of crime, police officers allegedly routinely reclassified crimes as less serious offenses and even discouraged citizens from reporting them in the first place.[35] CompStat shows that when data drives decisions, decisions about how to record the data will be distorted.

Still, data-driven management for cities is an irresistible fiscal force shaping the future. Ironically, this has been proven most starkly in Baltimore, the setting of *The Wire*, the critically acclaimed television series that lambasted the destructive and corrupting influence of CompStat-style management. Applying CompStat techniques to other aspects of government like trash collection and pothole repairs saved the city at least $100 million during Mayor Martin O'Malley's first term in office.[36] One former official puts the savings as high as a half-billion dollars for his entire administration, which ended in 2007.[37] Not bad for a system that cost just $20,000 to set up and $350,000 a year to run.[38]

Tolva's vision has a convincing air of inevitability. When I asked him to speculate on what big data means for cities in the future, his response was quick and terse. "Governing and policy making based on what the vital signs are telling us, not anecdote," he said.[39] Perhaps not surprisingly, his partner in reinventing Chicago's government as a data-driven enterprise is himself a crime mapper. The country's first municipal chief data officer, Brett Goldstein was brought over from the Chicago Police Department where, Tolva says, "he was crunching huge amounts of past crime data to nightly redeploy squads based on probability curves of incidents." But in his new role Goldstein can look beyond just police reports, at the many other socioeconomic indicators that can help suss out the conditions that foster crime.

Tolva believes it will take a culture change in city government to realize the full potential of bigger data and deeper analytics. "If you have a department that does it, you are probably doing it wrong and it is not suffused throughout. Success would be not needing a champion of data-driven decision making in the mayor's office," he says.[40] But it will take still more than culture change to use big data wisely. As it comes to inform more and more policy decisions, city leaders will have to become more sophisticated in how they evaluate data, whose indications are far more subtle than even the simple statistics they have relied on for many years. As Joe Flood found in his study of John Lindsay's administration in New York City in the 1960s, mayors often champion new data tools and methods without really understanding them. "I remember that I once wrote a speech for Lindsay and he made me use the phrase 'new budget science' three times in it," one budget aide told Flood, "and I'm convinced he didn't know what the words actually meant."[41]

"All Politics Is Local"

As we've seen, most cities that have sponsored apps contests skip over the most important step in design—identifying users' needs. In-house number crunchers like Tolva largely serve the needs of their peers, using technology to improve the effectiveness of government agencies. But as Tip O'Neill, the lion of Massachusetts politics, famously said, "All politics is local." Not surprisingly, in O'Neill's hometown of Boston, Mayor Tom Menino began building a smart city by using technology to hack new solutions to address the everyday problems of citizens. In 2010, he created a task force to rapidly prototype new civic technology, the Office of New Urban Mechanics.

Menino had a head start putting citizens first. If you want to start building a smart city by tackling problems rather than exploring what you can build with new tools, it certainly helps to have been in office for two decades. Over his long tenure, the mayor had built up a political nervous system that spanned the metropolis, pumping community concerns into his staff's BlackBerrys minute by minute. But Menino was on the beat himself—as he announced in early 2013 his plans to finally retire, a *Boston Globe* poll reaffirmed a widely known statistic: the mayor had personally met more than half of his constituents.[42] He didn't have to mine massive databases or launch apps contests to find new problems; his To-Do list was already a mile long. And with a deep understanding of where the gaps in the system lay, one could fine-tune instead of overhauling.

The Office of New Urban Mechanics' name conjures images of overall-clad technicians spelunking into grease-filled gearboxes deep under the Brutalist architectural abomination that serves as Boston City Hall.[43] But as Nigel Jacob, the group's cochair explained, it was a reference to Menino's own early life as data junkie. In the late 1980s, "He was a city councilman, and his vision of the city was focused on livability. He was entirely focused on classic quality of life indicators." This pragmatic focus on street-level performance led the *Boston Globe Magazine* to dub him in 1994, "the urban mechanic."[44] Unlike cities where the mayor's tech stars were busy launching apps contests, publishing open data, or running analytics, in Boston the mayor focused them on building tools for citizen engagement. "Technology is not part of our mission," explained Chris Osgood, a veteran civil servant who previously worked for New York City's Department of Parks and Recreation and who, as Jacob's cochair, made up the other half of the Office of New Urban Mechanics. "It is to connect people and government better."

Consider Boston's approach to the snow problem, as compared to Chicago or New York. Just as those cities were opening up their snowplow maps in January 2012, New Urban Mechanics launched "Adopt-A-Hydrant," a Web app that allowed neighborhood volunteers to claim local fireplugs as their own winter wards. On top of responding to over five thousand fires each year, the Boston Fire Department is responsible for shoveling out over thirteen thousand hydrants after every major snowstorm. In future snowstorms, Adopt-A-Hydrant will send text and e-mail alerts to let people know when and how to properly remove the snow from the fire hydrant they've "adopted."[45] It's an interesting, lean model, a low-tech approach that relies on citizens' own labor and the existing cellular network. It signalled that the city was doing something about snow without the need to spend lots of money. But was it practical? Scanning the site in summer 2012, six months after its January launch, I found fewer than a dozen hydrants claimed. I called up Jacob, who explained that the system hadn't yet been truly tested due to a lack of snow the previous winter. "We were a victim of global warming," he mused, "We never had a chance to use it."[46]

Compared to most of the cities we've seen, New Urban Mechanics was founded on a fundamentally different philosophy about how technology can be used to transform local government. For Osgood, the big opportunity was undoing decades of inward-looking thinking in city government, which he felt was being amplified by data-driven management and writing citizens out of the loop. "We've become so focused on, how much faster can we fill that pothole? How much quicker can we remove that graffiti?" he asked, lambasting the approach embedded in programs like CompStat, "that we try to quickly optimize our own operation in a way that actually doesn't engage constituents and make them part of the design process." Osgood continued, "Think of how Wikipedia was built. Think about how Google gets stronger every time somebody does a search. They've made it very simple for people to get involved in the process and strengthen the product with their own participation."[47]

The New Urban Mechanics team didn't just preach crowdsourcing; it also relied on crowdsourcing to get its own work done. While the office didn't have its own budget (it didn't officially exist in a formal bureaucratic sense, to keep it "lightweight" and preserve "a start-up character we have gone to great lengths to maintain," Jacob said), Jacob and Osgood and five program managers spread through various city departments formed a network funded by about $300,000 in city money and a slew of grants from local and national foundations.[48] It was a sizeable workforce for technology innovation compared to peer cities, and the team leveraged it to the hilt. "We don't try to do any of the work ourselves," Jacob explained. "We try to find people that are doing work in the space already that have similar goals to us, that we can partner with and actually deliver on." Instead of micromanaging, they stayed strategic. "We think about design, and we think about more classic policy questions." Procurement rules put in place to fight corrupt contracting limited its ability to quickly purchase new technology. But they embraced the $10,000 limit on what they could spend without a lengthy bidding process as a useful constraint that "forces us to think lean and mean," Jacob said. And it speeds things up. "You're talking about very small dollars. You're talking about weeks versus months."[49] Above all, "Urban Mechanics is an experimental laboratory," he told me.[50]

All of these factors—the focus on citizens, the substantial human resources, the severe constraints on project scope, the political reality that Menino doesn't have to grab headlines with every tech initiative—united to chart a markedly different path for Boston, an almost guerrilla approach to smart-city building. Like the minutemen of the Massachusetts rebellion, the New Urban Mechanics team picked its targets carefully, and struck fast with a tiny force. It's a point not lost on the team. Jacob saw early on that the contestants in city apps contests were "basically developing solutions for themselves. Which makes sense, right? Because that's how you scratch your itch." Boston chose not to follow that path. As Osgood saw it, Menino's focus on accountability to his constituents dictated a more engaged approach to apps. "Because of our mayor, we take very seriously the responsibility that government has to understand the problems that residents have, and to try and solve those particular problems." Ensuring that the apps New Urban Mechanics built were both useful to Boston residents and "piloting something interesting and creative" perhaps results in fewer apps, he says, but apps that will be "sustained and evolved and resonate more."[51]

Unlike other cities, where technology is seen as the catalyst of change, Menino made technology subservient. Although it's a unique creation of his long tenure and style of governance, Boston's strategy could be the most universally viable model for civic technology out there. It's the first approach to smart cities that feels as though it was designed by a political scientist rather than a software engineer. It is subtle and measured where others are bombastic about the benefits of technology. Jacob's assessment is telling of the team's cautious approach to tinkering with the relationship between government, people, and innovation. "I think in general cities have only a very weak understanding of what people need or how technology could be used to address social problems," he concluded.[52]

Boston's approach of guiding technological innovation with smart politics has caught the attention of mayors elsewhere. As Jacob explained to me later, in August 2012 he had taken on a new role advising his peers in several other American cities on how to replicate the success of the Office of New Urban Mechanics. Philadelphia, the first to come knocking "actually called and asked 'Can we just franchise what you guys do?'" Jacob proudly said.[53] He was also working to help spread to other cities some of the projects kick-started in Boston. One such tool, Community PlanIt, was an online game designed by Eric Gordon, a visual and media arts professor at Emerson College, to enhance the value of community meetings. When we spoke, Community PlanIt had been successfully rolled out in two of Boston's suburbs as well as Detroit.

Although it was poised to go viral, can New Urban Mechanics survive a change of leadership at home in Boston? Menino will finally leave office after the 2013 mayoral election, having served a record five terms. Both Jacob and Osgood believed that their approach already had the critical buy-in from citizens that eluded efforts in other cities. As Jacob saw it, "There [has] definitely been a problem . . . with some cities where the focus of innovation is about business process and improvement. Those are easy things to cut at budget time. It's very hard to argue against a program that has been rolled out to the constituents . . . especially if it's something that is successful and the people are engaged." For Osgood, engagement may even be what really matters in the end, more than any particular innovation itself—the novelty of the new technology "should be a distant second, relative to improving new models of civic engagement or adding value to the lives of constituents."[54]

For Jacob, technology had opened the door for change driven from, but mostly happening outside, City Hall. "When I imagine the city operating on a different model," he opined, "I think that people will be empowered to do things that, right now, are done exclusively by government. We would need to rethink a lot of the traditional roles. . . . People need to be able to see a way to make life better for themselves, as opposed to waiting for government or for some magical start-up to do it for them."[55]

Betting the Farm On Smart

So far, American forays into building smart cities have been spasmodic, on-again/off-again affairs. But in Spain, with an economy in free fall, the city of Zaragoza is completely reinventing its physical landscape, its economy, and its government with smart technology.

"There is the antenna," Daniel Sarasa says, pointing. A tiny white plastic bud juts from a street lamp, just beyond the bust of Spanish painter Francisco Goya that dominates this end of Plaza del Pilar, Zaragoza's central square. He steps out of a long winter shadow cast by of the looming basilica cathedral, an austere block of Iberian stone. "The whole plaza was filled with tents."[56]

Months before American cities faced the "99 percent" during the fall 2011 Occupy protests, Spain erupted in dissent. Plaza del Pilar was the epicenter of the "15-M" movement (for May 15, the day the protests started) in Spain's fifth largest city. At their peak some ten thousand people gathered here to demonstrate against austerity measures taken by the Spanish government as the country struggled to stabilize its debt and stay in the good graces of international bond markets.

In the United States, Occupy encampments used cellular networks to keep organizers online, but in Zaragoza a new public Wi-Fi network, years in the making, was coming online just as the protests swelled. One of the mayor's key digital strategists, Sarasa explains that the antenna in Plaza del Pilar was just one of a cluster installed earlier that spring throughout the city at locations suggested by citizens in a public survey the year before. As May 15 approached, the network was going through final beta tests and had not yet been formally launched. But protesters quickly discovered the service and logged on in droves, bringing transfer speeds to a crawl. Conspiracy theories of a city-ordered shutdown swept across Twitter, much to Sarasa's surprise. "I tweeted, telling them about other nearby hot spots, and urging them to go there to connect." In American cities it was police, often outfitted in riot gear, who dealt with protesters that year. But in Zaragoza, city agencies instead used social networks to shepherd them in peaceful digital dissent across an archipelago of wireless hot spots.

When fully built out Zaragoza's Wi-Fi network will involve over two hundred hot spots blanketing a zone dubbed the Digital Mile (or Milla Digital, in Spanish). The Digital Mile stretches from the Plaza del Pilar at the city's center to the site of the 2008 World Expo across the Ebro River, a riverfront to which the city had long turned a cold shoulder. The path is a microcosm of the city's journey through history. At one end is the basilica of Our Lady of the Pillar (Nuestra Señora del Pilar), the cathedral whose construction from 1681 to the middle of the twentieth century, when its towers were finally completed, coincided with Spain's decline from global empire to shattered, war-weary backwater. At the other end, the Expo site had served briefly in 2008 as a venue for reimagining the city's future.

The Digital Mile is the centerpiece of a broad effort to turn Zaragoza into what Sarasa describes as an "open-source city." "We had to come up with something new," he explains. While Zaragoza occupies a strategic redoubt on the road from the political and economic capital, Madrid, and the resurgent seaside cultural entrepôt of Barcelona, it lives in the shadow of both. "When we started out, we knew this wasn't Madrid. And there's no beach. Woody Allen isn't coming here to make movies," he says, referring to the director's 2008 hit *Vicky Cristina Barcelona*, shot on location in that city. If it were to be anything more than a provincial hub, Zaragoza had to do something radical. As the city worked with a group of MIT urban design professors, plans for the Digital Mile quickly took shape. The Media Lab's William

Mitchell, author of several books on cities and digital technology, teamed up with Dennis Frenchman, the head of MIT's urban design program. Frenchman had previously crafted designs for smart streets in South Korea, England, and Abu Dhabi that shrewdly deployed new digital technologies to enhance the vitality of public places. For instance, in Seoul's Digital Media City, a predecessor to Songdo, Frenchman designed a series of multistory screens that would stretch in an unbroken line down the site's main, pedestrianized "media street." It was akin to Times Square's brilliant signage, but instead of a dizzying jumble of ads, the entire system could be operated as a single screen to display artwork, celebratory images, or, in an emergency, evacuation instructions.[57] For Zaragoza he proposed a necklace of new buildings and public technology exhibitions that would similarly weave connections between the digital and physical city.

My tour of Zaragoza had begun earlier that morning with Juan Pradas, one of Sarasa's colleagues, at the center of the Digital Mile, where Zaragoza has literally put itself back on the map. At a massive new rail station, bigger than most airport terminals, sleek new bullet trains slide to a stop before whisking passengers off to Barcelona and Madrid, less than two hours in either direction.

The station has sparked a miniature building boom. Traversing a delicate pedestrian bridge designed by Frenchman to overstep a mid-twentieth-century traffic circle that couldn't be moved, we approach a trio of sleek new buildings clad in frosted white glass. The two larger ones will house the Center for Art and Technology—"the CAT" in Pradas's jargon. It is a spitting image of the Media Lab's new building in Cambridge, and it was the last great dream of Mitchell, who had passed away a year earlier. The resemblance is more than cosmetic, for the CAT is also destined to become the kind of place where artists, technologists, and citizens come together to explore the possibilities of smart technologies to reshape the city. It is to be, as Michael Joroff—another MIT advisor to Zaragoza has told me—not merely a think tank but a "think-do tank." The hope is that it will be a source of bottom-up innovations, an open-source department of civic works. The smallest of the three buildings, a business incubator, is already open. We peek inside, and the pleasant hum of digital designing fills the air—a buzzing espresso machine, electronica beats, and fingers tickling keyboards.

Water defines the Digital Mile: it was the theme of the 2008 Expo and is a precious resource in the city's arid region. Moving on from the CAT, we explore a network of technology-studded public spaces, including MIT professor Carlo Ratti's Digital Water Pavilion, which encourages people to interact with and even program smart systems. A fountain that works like an ink-jet printer, the Pavilion sports two long lines of water cannons that shoot sheets of liquid down from an overhead canopy. As you bravely leap through, a sensor catches you and magically cuts off the flow, creating a human-sized safe haven. After you pass through, the watery wall closes behind you.

More important, though, the Pavilion is a literal interpretation of the idea of an open-source city, with multiple layers of programmability. Amateur hackers can send a text message to the controls, directing the jets to fire in sequences that write your message in patterns of falling drops. A few pecks on my phone and I'm programming the streetscape of Zaragoza. For the pros, there's an API for coding apps that add new behaviors to the fountain.

More digital waterworks are planned. Sarasa describes plans for the Digital Diamond, a public swimming pool proposed for a nearby residential area, that he hopes will be warmed on the region's cold desert nights by the waste heat from a nearby server farm. Across the river lies the empty Expo site, stuck, as those kinds of places always are, in limbo as the city tries to figure out how to best reuse it.

Backtracking past the CAT, the Digital Mile winds its way through the existing city on its way to Plaza del Pilar. We cross over into La Almozara, a high-rise block built atop the former site of a chemical factory, now home to a large community of working-class Romanian immigrants. At its center, I find more Wi-Fi hot spots, clustered around the neighborhood's Centro Civico community center. A utilitarian relic of Spain's post-Franco socialist renaissance, the boxy brick low-rise sits in a small plaza surrounded by ten-story apartment buildings. Zaragoza is upgrading these community centers for the twenty-first century. One side effect of the Wi-Fi project was that it created an excuse to run fiber-optic lines to all seventeen Centros Civicos throughout the city. The guard at the front desk, no doubt himself a member of the left wing's old guard, turns to open a cabinet and reveals a twinkling array of Cisco routers.

More than the fastest Wi-Fi, the biggest new tech center, or the entire Digital Mile, the humble "citizen card," issued under a new city initiative, is already transforming Zaragoza. The cards are only available to residents—migrants from other cities who don't register with authorities can't get one. But, I suppose, even in a smart city, you can bend the rules on occasion. Pradas beckons me outside to a rack of public bicycles just outside and taps a card to unlock my ride. I offer some euros, but he shrugs and smiles. "It's OK, it's my daughter's card."

A stunningly simple innovation for a world of face recognition and predictive modeling, the citizen card is a key that unlocks Zaragoza both online and in the real world. The same card that unlocks a bike share will get you on the Wi-Fi, check out your books at the library, and pay for the bus ride home. Shops and cafés offer cardholders discounts, which has made the program wildly successful—over 20 percent of the city's 750,000 residents signed up in the first year. As Sarasa explains, "This is all about engagement. . . ." Pradas cuts him off, pronouncing with certainty, "the card creates a sense of belonging."

The citizen card promises to fundamentally change how the city works. There are plans to create a kind of game, a frequent-user program that offers "digital miles" as rewards to heavy users of the bus system and Wi-Fi network. "The card generates a lot of data on activities, and is a powerful tool for planning," Sarasa points out. Patterns in card use allow city managers to see how people use public services in great detail, allowing those services to be managed in a more holistic way. Unlike Global Pulse's contortions to anonymize and obscure individuals' data, Sarasa sees the city as the best possible referee in a world of urban sensing. "There is a Big Brother aspect we are aware of. But we think the City can be a very good keeper of citizens' privacy." Given the hand-wringing debates around the proliferation of individually identifiable data online, and the near-total lack of good ideas about how to deal with it, the idea of local governments as custodians of our personal data is intriguing to me. Is it a power grab by government or inspired leadership? My feeling leans to the latter. But the thought of American cities stepping into this role seems, sadly, unlikely given the enormous responsibility it would entail.

Zaragoza certainly is one to beat in the emerging world of smart cities. Its physical transformation has been bold but carefully measured. It is building world-class facilities that will enable smart-city innovation and economic growth in the future, but has balanced it with upgrades to community centers and public spaces. The citizen card has enormous potential to change the nature of citizenship. None of these pieces alone is a silver bullet. But together they are a "platform for innovation," as Sarasa describes it. This is no company town rising in an open field, an enclave of iPhone-toting hipsters, or a bid for headlines as an election approaches. It's a real city, with real problems, thinking and investing long-term in the most promising set of tools at hand.

For all its promise, Zaragoza has a rough road ahead. As the Digital Mile moved into the second half of its first decade, Spain's economic crisis went from bad to worse. The outlook was more dire than at any time since the nation's devastating Civil War in the 1930s. Overall unemployment hovered around 25 percent. For those under twenty-five years of age, the Digital Mile's future caretakers, it surged past 50 percent and underscored their angry 2011 occupation.

Spain's economic troubles have turned the Center for Art and Technology into a rallying point for the civic and business leadership of Zaragoza. As Pradas explained to me, before the crisis local business leaders barely paid attention to the project. But as the opening in summer 2012 approached, his phone was ringing off the hook with offers of assistance. But building support among young people will be far more difficult. In the past, open-source hacker groups and free-wireless cooperatives had built working relationships with the city. For instance, "Cachirulo Valley," a colorful group jovially named after a kind of knitted scarf worn in the region, holds its meetings in a conference room carved out of the basement of the Digital Water Pavilion. But a new crop of movements, formed by the May 15 protests, have refused to deal with government. Pradas sees the Center, which will be run by an independent foundation, as a possible neutral ground to bring the parties together.

The stakes in this gamble on smart couldn't be higher. Built with the last of the national government's massive 2009 stimulus spending, and opening in the wake of a series of elections that saw Mayor Juan Alberto Belloch barely survive a nationwide left-wing rout, the outcome of Zaragoza's gamble on the Center for Art and Technology campus is far from certain. But it is an inspiring example of how different a smart city can emerge when civic leaders craft a big vision that reflects the needs and aspirations of an entire community and mobilize the resources to deliver it. Corporate smart cities chase the Holy Grail of efficiency, while the grass roots explores the possibilities of social technology, but in Zaragoza—as in most real cities today—citizen engagement and economic development are the pressing challenges.

Sarasa makes the priorities clear. "We are creating a machine to create jobs. This has to produce things for the city."

Leadership for the Smart City

Toward the end of 2011, IBM threw a big party for Rio de Janeiro. "How is that mayors are getting things done, while other leaders seem stuck?" asked IBM's chairman Sam Palmisano from the stage of the Smarter Cities Forum. Palmisano would soon step down, easing into retirement after setting the firm on course for decades' worth of smart-city–driven growth. Rio's mayor, Eduardo Paes, basked in the appreciation, quietly calculating the political payout from his decision to call in Big Blue to build the city's Operations Center. "These city leaders are nonideological," Palmisano posited, echoing storied New York City Mayor Fiorello LaGuardia, who famously said "There is no Democratic or Republican way of cleaning the streets." Palmisano concluded, "They get things done. . . . Smarter city leaders think—and manage—for the long term."[58] At the very least, for the next election.

Today, cities are the most pragmatic and effective level of government. In an era of gridlock at the national level, as Parag Khanna and David Skilling, who both serve as foreign policy advisors to the nation-state of Singapore, have argued in their essay "Big Ideas from Small Places," "cities and provinces around the world are assuming a more important leadership role on global policy issues."[59] Even as they grow larger, cities maintain a sense of shared destiny that mobilizes people to work together. We've seen this pragmatic focus at work throughout this chapter. Consider what the average mayor sees when she looks out her window at the city she runs—the thankless work of delivering reliable transportation, safe streets, and high-quality health care and education. These functions dominate municipal budgets and draw public ire when they are mismanaged. As they grasp for solutions, these new technologies hold tremendous appeal.

Companies like IBM think the solutions they've built for business can solve problems for city governments. But cities aren't companies. Big technology companies have spent a half-decade educating mayors about technology, yet their own understanding of how cities work is wanting. As Boston's Nigel Jacob explained, "We've seen nothing but missteps" from industry. "Because they want to see the city broadly as an enterprise, they make a huge number of assumptions about what's driving what. They will often miss huge dimensions of how we actually operate."[60] Palmisano may be able to convince himself the world's mayors are nonideological technocrats, but his employees and their customers face the reality of urban politics every day.

The limits of grassroots methods of engineering smart solutions to public problems are clear too. Few civic hackers want to do the dull, dirty, and dangerous work that IBM's engineers are asked to tackle. For all their creativity, apps contests still haven't produced much of lasting value for the broader public. One of France's leading Internet activists, Daniel Kaplan, said it best when he called the results of apps contests "mostly proofs of concepts (or of their programmers' skills), with at best anecdotal benefits for ordinary citizens."[61] Efforts to mobilize citizen participation through the Web and social media have their work cut out for them.

So who is going to design the smart city of the future, if the geeks on both sides of the street don't truly grok the challenge? In the end it will be up to the mayors and their teams. They'll hedge their bets, buying things from big corporations while simultaneously seeding grassroots efforts to solve the same challenges. When that doesn't work, they'll just build their own. They'll do whatever it takes to get the job done with the limited resources they have.

But as a new model for civic leadership in the design of smart cities unfolds, there are lots of open questions. How can they make sure there are opportunities for both industry and grassroots efforts to innovate? How do they empower citizens to create and provide new public services but not be tempted to offload government's responsibilities? How will they collect and aggregate data about the city for public good but build safeguards to prevent it from being misused? None of these issues will be resolved soon, and they will crop up repeatedly. One thing is sure. They will land on the desks of city leaders, because no one else will want to deal with them.

8

A Planet of Civic Laboratories

Peter Hirshberg spins his laptop around. Bold white letters on a black background spell out *OccuAPI.* "I have no idea what it means," he chuckles. "But I like it."[1] It's November 2011. A dozen blocks south, the Occupy Wall Street protests are reaching their violent zenith in Zuccotti Park. The city is on edge from daily marches that take the "99 percent" and their riot gear–clad chaperones tramping across Manhattan. Police helicopters hover like angry wasps overhead. Hirshberg's neologism is an attempt to capture the excitement of the Occupy movement as well as the more subtle technological transformation of citizen-government interaction by open data and apps.

America is no stranger to youth movements, though it had been a long time since one loomed so large in the public mind. The closest analogue is probably 1967, when tens of thousands of young people descended on San Francisco's Haight-Ashbury district. In a hothouse of social experimentation that became known as the "Summer of Love," they shared everything—housing, food, drugs, and sex. The enormous cultural impact of that psychedelic freak-out on American society can be felt today, and it still casts a long shadow over San Francisco. There, Hirshberg has been a driving force behind a new creative space just down the hill from Haight-Ashbury, the Gray

Area Foundation for the Arts. Both physically and spiritually, it sits at the intersection of that 1960s counterculture and a new techno-utopianism. It's just a few steps to either Twitter's headquarters or the head office of Burning Man, the radical art festival that builds a temporary city in the Nevada desert each summer.

Though he takes inspiration from the hippies, Hirshberg is politically pragmatic. He soon slaps his laptop shut and stops playing dumb. "Look," he says, "in the 60s you protested the establishment. Today you just write to its API." For Hirshberg, the way to accelerate change is to plug revolutionary software directly into government databases.

Nowhere has the creative urge of smart-city hackers come into such direct synergy with efforts to reinvent city government than in San Francisco. The story begins in November 2010, when longtime mayor Gavin Newsom was elected as California's next lieutenant governor. With more than a dozen candidates tossing their hat in the ring to succeed him at the city's helm, and the local economy once again riding a frothy wave of start-up-driven innovation, Hirshberg saw an opportunity to spark a public debate about how technology could be harnessed to improve government. He first tried to convene a workshop with the candidates, but was overcome by what he describes as "enthusiastic data syndrome." Things didn't click. It was "the classic conversation the geek has with the business user," Hirshberg says.[2] The candidates didn't get it.

Evoking a left-wing hero of the 1960s, Abbie Hoffman, whose unforgettable *Steal This Book* was a foundational text for the Youth International Party (the "Yippies"), Hirshberg explains how he hacked the election. "I realized, all we need to do this summer is come up with ideas worth stealing. We need the political class to see this as a form of innovation." More than four decades after the Summer of Love, in 2011 he proposed a Summer of Smart. An epic civic hackathon, Summer of Smart was designed to engage the candidates and their constituents around tangible tools, rather than abstract concepts like open data. Instead of asking for resources, they would turn the tables on candidates and offer up solutions. San Francisco would once again become a social laboratory. But this time, peoples' minds would be opened not by LSD but by the wonders of information technology.

The next step was getting people involved. Hirshberg knew how to enlist techies, artists, and activists—the Gray Area Foundation already had an impressive community around it. But he needed to plug government in. Apps contests in other cities had been organized by government, which maintained an arm's length relationship with the contestants. Aside from sharing data, there was no real collaboration between government and citizens. So Hirshberg reached out to Jay Nath, the city's director of innovation. An up-and-comer at City Hall, Nath had recently pushed through the nation's first municipal open-data legislation. Instead of haphazardly releasing data for apps contests at the mayor's behest, San Francisco's agencies were now required by law to systematically share as much as could be done safely and legally.

But even with such progressive legislation, the city was sitting on a massive stockpile of unreleased data. By Nath's estimate, there were tens of thousands of databases hiding in the city's servers, including a ten-year digital record of over a million police reports. Nath wanted to find more ways to get data into the hands of people who could create valuable services with it. "The city is a monopoly. We are stewards of the data. This is data that belongs in the public domain," he said.[3]

Openness was already paying off for Nath. When he joined the city years before, he had overseen a budget of millions and a staff of twelve working on the city's 311 system. Working with OpenPlans, a New York–based nonprofit, he had launched an open 311 system in March 2010. For the first time, it was possible for anyone to create apps that could send data back upstream to the city's computers—noise complaints, service requests, pothole reports.

The new system held the potential for a vastly expanded, bidirectional flow of timely information between citizens and government, much as Hirshberg had envisioned. By the summer of 2011, budget cutbacks had reduced Nath's staff to two. But by expanding access to this data, he explained, "I was actually getting more done."

Summer of Smart came to a head during three summer weekends in a series of hackathons that Hirshberg recalls as "electric." Starting on Friday at 5:00 p.m. with an inspirational talk, each dealt with a different area of city life. The first focused on community development and public art; the second on sustainability, energy, and transportation; and the third on public health, food, and nutrition. Over the course of the summer some five hundred hardware hackers, software developers, students, artists, designers, and community activists put in over ten thousand hours of volunteer time to create twenty-three interactive projects.[4]

Unlike previous city-sponsored apps contests, Summer of Smart's success stemmed from its laser-sharp focus on problems and its intense face-to-face teamwork by a broad swath of stakeholders. Nath recruited the front-line managers who run the city's transportation, housing, and schools day-to-day so that people with firsthand experience with the challenges of government could help steer the work of the hackers. Hirshberg recounts how one discussion around fixing the city's slow and unreliable Muni transit system turned into an ad hoc visit to the nearby control center. The outing thrilled the digital trainspotters who had given up their weekend to help the city, but more importantly, it showed them the real capabilities and constraints public managers face every day. The intensity of the events pushed people to focus and collaborate. "Fast prototyping was what got the partners to engage each other," says Hirshberg.[5] The participation of the mayoral candidates—who all dropped in—tantalized volunteers with the prospect of real civic impact.

Some compelling apps emerged from Summer of Smart. GOODBUILDINGS mashed up city records with related information from across the web, like walkability scores, to guide people seeking commercial space in sustainable buildings. Another app, Market Guardians, used game mechanics—awarding virtual points and badges to the most active participants—to entice young people to map urban "food deserts" by tracking the availability of healthy food at stores in inner-city neighborhoods. In October, the winning teams presented their projects at a mayoral candidates' forum just three weeks before the election. Nath hammered the message home, telling his colleagues in government, "the community isn't just a way to define, but also a way to solve problems."[6]

In 2012, with Hirshberg's protégé Jake Levitas now at the helm of its civic hacking efforts, the Gray Area Foundation began to refine and export its model for civic engagement around smart technology, launching what it now called "Urban Prototyping" events in San Francisco and Singapore. Next came London in early 2013, with potentially dozens more events around the planet to come. Whereas Summer of Smart's key innovation was its intensity and participation of nontechies, Urban Prototyping raised the stakes by focusing on quality and sustainability. The process began with an open call for projects that combined digital and physical elements of the city, especially open-source designs that could be readily replicated in other places. In San Francisco, over a hundred proposals were submitted; eighteen were selected. They received up to $1,000 in funding, a workspace, technical assistance from Levitas's group, and support from the city to deploy their prototypes along a street in San Francisco's mid-Market neighborhood. Reliving the Summer of Smart, the teams gathered for a weekend "Makeathon" to bring their designs to life.[7]

Summer of Smart was itself a clever hack, ushering a marginal movement from the geek fringe to the center of civic debate. More importantly, it established a new model for government and citizens to work together to use technology to address pressing needs. San Francisco has shown that it won't simply install shrink-wrapped software dreamed up in corporate labs. It will be a smart city that thinks for itself, a permissive place to prototype the future.

Places That Make Software

San Francisco is just one of thousands of civic laboratories, innovative communities where people are eagerly adapting smart technology to unique local needs. This is a strange development for a world where multinational corporations have become so adept at standardizing and spreading new innovations. As we have seen in earlier chapters, companies like IBM and Cisco would love to do the same with smart technologies for cities. In the August 2011 issue of *Scientific American*, MIT's Carlo Ratti and I published an article celebrating the groundswell of design innovation in these pioneeering communities. On the back cover, an IBM advertisement issued a terse rebuttal: "A smart solution in one city can work in any other city." It sounded like a proposal to mass-produce urban intelligence.

The beauty of cities is that no two are precisely the same. Each has a unique history, architecture, politics, and culture. Even the smallest town is a collection of households who have over the years built up a shared identity and arrangements for working, living, and playing together. New communities differentiate in this way astonishingly fast, typically in a generation or less. In the 1950s, Long Island's Levittown was the poster child for homogeneous, mass-produced American suburbia. Driving through today, you can hardly find any two houses that still look alike. Over the last half-century since they were built, they've been expanded and customized by their owners in countless ways. By living together in our cities, tweaking their basic design to meet our changing realities and forging social bonds with our neighbors, we make them uniquely ours. That's why urban design is as much an art as it is science. It has to respond to countless local variables and idiosyncrasies.

While it hasn't always been that way, today, technology design is becoming more like urban design. For the last century, our devices were highly standardized objects, produced identically on an industrial scale, and designed to perform a few functions. As late as the mid-1990s, in the course of a year you probably only used a single computer—usually a desktop model—and a handful of software packages. Today, we routinely "do" computing each day though conscious and unconscious interactions with dozens or even hundreds of different kinds of devices running thousands of different pieces of software—our laptops, iPads, and smartphones to be sure, but also computers embedded in buildings, appliances, automobiles, traffic signals, and so on. Mobile devices have liberated computing from the desktop and kicked this shift into high gear. Now digital technology has to respond to and engage with what's happening around it, just as good architecture requires careful consideration of a building's surroundings.

In 2004, social-media guru Clay Shirky gave a name to the kind of technology created by place-based communities: "situated software."[8] Years before Apple launched its App Store, Shirky noticed that his students at New York University's Interactive Telecommunications Program were building social software for themselves using nothing but open-source code and microcontrollers. Their approach was antithetical to the "Web School" that had prevailed up to that point, "where scalability, generality, and completeness were the key virtues." Instead, situated software was "designed for use by a specific social group, rather than for a generic set of 'users.' "[9]

You can find situated software on any smartphone, where for almost any life situation one might encounter, as Apple's ads proclaimed in 2009, "There's an app for that." Some apps are only for use on the go. Others are for certain kinds of places, or specific social settings. For instance, iTrans will give you the schedule for the subway into Manhattan, and Exit Strategy will tell you which car to ride so you're closest to the correct egress when you disembark. It also cleverly caches a street map of Manhattan that you can browse offline underground, because New York is alone among world cities in its lack of underground mobile coverage. In San Francisco, Uber can summon a taxi with one click. In Manhattan most us still hail our cabs by hand. But in a pinch you might reach for CabSense, an app that analyzes millions of location-tagged taxicab pickup records collected by the city to identify the best corner to catch one. In Tel Aviv there's an app that sends alerts whenever Hamas rockets are inbound from the Gaza Strip. An alumnus of the MIT Media Lab living in the Iraqi capital of Baghdad has reportedly designed an app that lists recent kidnappings and the going rate for ransom. And Apple's Siri, which hails from Silicon Valley, might be the most suburban technology ever created: its voice recognition is perfect for connected cars but completely useless on noisy city sidewalks.

Shirky's students built situated software because, for the first time, they could. "Making form-fit software for a small group of users has typically been the province of banks and research labs," Shirky explained. "The kinds of scarcities the Web School was meant to address—the expense of adequate hardware, the rarity of programming talent, and the sparse distribution of potential users—are no longer the constraints they once were."[11] Today, the infrastructure that's needed to build and distribute a smartphone app is already in place, and either free or rentable; it costs almost nothing to turn a novel idea about how to interact with the city into a piece of software that meets the needs of a handful of people in close proximity. For Shirky, situated software didn't even have to be that good, as long as it scratched some collective itch.

Situated software also connected the Web to the physical world. In fact, those connections were critical to making the designs successful. The two student projects that inspired Shirky's thinking, Scout and CoDeck, both "had the classic problem of notification—getting a user to tune in requires interrupting their current activity." Both "hit on the same solution: take most of the interface off the PC's dislocated screen, and move it into a physical object in the lounge, the meeting place/dining room/foosball emporium in the center of the ITP floor." Scout was like Dodgeball, which we saw in chapter 4, but instead of using phones to check in, students swiped their university ID card. CoDeck was basically YouTube wrapped inside a 1970s Betamax videocassette player: people could use its buttons as controls to share and comment on each other's video creations. Unlike our experience of software on the desktop, where the entire experience plays out in a single window, situated software spills out into the larger world and inserts itself into our lives. Both projects had websites where users could interact with the software, but as Shirky noted, "the core piece of each is location in physical space that puts the application in a social context."

Shirky's essay was a powerful premonition of how the smartphone software ecosystem would unfold. As the same conditions that had existed inside ITP were duplicated in entire cities—wide adoption of smartphones, heavy use of online social networks, and a sensory infrastructure that phones could use to orient themselves to the physical world—demand for situated software exploded. When Apple's iTunes App Store opened, it put an odd twist on Shirky's original model by making it possible for situated software to exploit a Web-scale distribution channel. By 2010 nearly one in three adult mobile phone users in the United States had downloaded at least one of more than five hundred thousand different apps available for smartphones.[12]

Now that it's on the street, computing will never be the same. The screens of our desktop operating systems like Windows and OSX are like the suburbs, split up into a handful of single-use zones—Microsoft Office, the Web browser, and deeply immersive games. The software ecosystem of the iPhone is instead a mirror image of the urban world it has grown up in—like a great city street, it's populated by quirky little storefronts that work together to create a fine-grained mix. The iPhone may have come from Cupertino, in suburban Silicon Valley, but its true potential is being realized on the streets of San Francisco, New York, London, and Shanghai.

Shirky's essay echoes Jane Jacobs's observations on great cities. "Situated software isn't a technological strategy," he writes, "so much as an attitude about closeness of fit between software and its group of users, and a refusal to embrace scale, generality or completeness as unqualified virtues." The grassroots revolution that transformed urban planning in Jacobs's era took on similar assumptions when it came to city design. It was a response to the excesses of urban planning's own "Web School," the large-scale reshaping of the city practiced by power brokers like Robert Moses with little regard for the street life of the city.

But for all his enthusiasm, Shirky was deeply skeptical of situated software's ability to scale beyond small social groups like his students. "By relying on existing social fabric"—the casual face-to-face encounters with fellow users—"situated software is guaranteed not to work at the scale Web School apps do." Situated software, by definition, needed an element of face-to-face interaction among its users.

But as we are seeing in these burgeoning civic laboratories, the scale of the city is an interesting intermediate scale at which many kinds of situated software can succeed. Beyond the intimate realm Shirky observed his students sharing, there are lots of shared contexts at the city level that aren't shared at the scale of the whole Web. Transit systems, with all their quirks, are distinctly different between cities and have spawned a whole category of situated software—developers in Portland, Oregon, a city of just 590,000 people, have created fifty apps for the region's transit system, each with its own unique package of features.[13] Climate is another trait that is relatively uniform at the city scale but distinguishes one city from another. (San Francisco, with its extensive range of microclimates, is an outlier here). All of these local variations are starting points for situated software. Apps for pedestrians, for instance, will have to understand differences in street culture. New Yorkers are chronic jaywalkers, but in Seattle people wait obediently at the corner for the signal to change.

It should come as no surprise that our civic laboratories are spinning out their own situated software. In fact, it would be weird if they didn't. "For if each human individuality be unique, how much more must that of every city?" asked Patrick Geddes.[14] The same urge that drives communities to differentiate themselves through physical design, regulation, and social norms will shape the way smart technologies are used to retrofit them. It's a mistake to assume that everything could or should be copied from city to city, however commercially attractive that may be. There are economies of scale, but there are also big benefits to doing it your own way. At the scale of big cities, these tradeoffs tend to be in balance.

"Build Locally, Spread Nationally"

Good ideas about smart technology are indeed spreading from city to city, but not quite the way IBM envisions. Rather, it's happening peer-to-peer, driven by a new crop of NGOs working nationally and internationally to cross-fertilize innovations.

For much of the history of cities, good ideas about how to design and govern them have spread slowly. As recently as the nineteenth century, if you wanted to spread a new idea in city planning, the best way to do it was colonization. The Romans laid down the basic template for much of urban Europe. British-trained engineers designed the flawless Hong Kong metro system (prior to the 1997 return to Chinese rule) by tapping a century's worth of knowledge gained building the London Underground. But, as we saw in chapter 3, professional urban planning and the peaceful and systematic exchange of best practices between cities is barely a hundred years old.

Recently, this flow has gone global. Rather than just borrow ideas from neighboring communities or national leaders, innovations are crossing borders at an increasing clip. Bus rapid transit, which combines curbside payment to expedite boarding with dedicated lanes to bypass traffic, began in Brazil in the 1970s but in the last decade has been implemented in Europe, Asia, and North America. Public bicycle sharing schemes have spread even faster, popping up all over the world after the launch of Paris's massive Vélib system in 2007.

Cheap air travel and the Web have been key to spreading these ideas. Watching a video of a huge crowd board a bus in seconds communicates the power of the idea faster than a pile of studies. Hearing the mayor of another city explain how he convinced voters to go along with the scheme is indispensable knowledge when you launch your own campaign at home.

So what happens when some hackathon or city agency comes up with a smart-technology idea that could work elsewhere? One organization, Code for America, wants to play Johnny Appleseed. "While each city has its own character and personality," writes founder Jennifer Pahlka, "at their core there are common needs, which can be addressed with shared and reusable solutions. In this age of shrinking budgets and rising needs, each city acting in isolation is no longer sustainable." The group's mission, in her words—"Build locally, spread nationally."[15]

Code for America actually started as an idea about how to fix the national government, but its proponents soon found that it worked better when scaled down to the local level. In 2008 Pahlka was running O'Reilly Media's annual Web 2.0 conference. She was a super-connected node in the tech community, and after the presidential election that year, she noticed that people in the tech industry were being tapped for transition team spots for the new administration. "It was clear that there was going to be an opportunity to do something with technology in the federal government that hadn't been possible earlier," she says. With Tim O'Reilly, a publisher of technical books and the tireless open-source advocate who coined the term "Web 2.0," she launched a new conference, Gov 2.0, to "bring the principles and values of the web to government."[16]

At first, Gov 2.0 had nothing to do with cities. The tech community, energized by the Obama campaign's promise of systemic change, was focused on transformation at the federal level. But as Pahlka filled her Twitter feed with thoughts and news about government technology, it caught the attention of Andrew Greenhill, chief of staff for the mayor of Tucson, Arizona, and the husband of one of Pahlka's childhood friends. As she recalls, Greenhill implored her by e-mail to help, wanting to know how he could entice developers to come to

Tucson and write apps for the city. Puzzled and frustrated, Pahlka said she wrote back, "I don't know. I can't help you."[17]

Greenhill continued to bug her. He called her, castigating Gov 2.0's focus on the federal government. Cities were facing "a huge financial crisis no one's talking about," Pahlka recalls him saying. Property values were falling, cutting into tax revenues, and cuts in consumer spending had hit sales-tax receipts hard. Pension funds were taking a huge hit just as boomers were lining up to retire. States, facing their own fiscal disaster, were rapidly cutting aid to cities. In a period of retrenchment, apps were a rare opportunity to innovate without spending a lot.

Cities also offered a chance for a visible and tangible impact on citizens' lives. "The federal government is so far removed from what actually happens to people in their daily lives," Pahlka explained to me. "But if your pothole gets fixed quicker, you notice. If your mayor is more responsive, you notice. If you're able to have an impact on your city's budget, you feel that. That was compelling to me."

Code for America was born, of all places, over a beer at an Arizona barbecue. In the summer of 2009, on a family vacation in Flagstaff, Pahlka's debate with Greenhill finally came to a head. "Andrew had done Teach for America, and we were talking about its impact and whether it was a good experience for him," she recalled. "We were talking about how people will do things that aren't money-driven in order to give back." The thread had come full circle when Greenhill asked yet again for help writing apps. "We need a Teach for America for geeks!" she blurted out.

Pahlka was electrified. "That night, I said to my dad and stepmom who were there with me, 'I'm going to quit my job and go to start this thing.'" Returning to San Francisco, she raised $20,000 from the Sunlight Foundation and the Case Foundation while producing one last Web 2.0 conference that fall in New York. In December she tendered her letter of resignation, and on January 1, 2010, Code for America began accepting applications for its first fellowship program.

Code for America solves a maddening problem that cities everywhere face when they try to institutionalize the guerrilla innovation methods of civic hackers. The projects come together too fast and are too small to fit into the painstaking procurement process that governs public contracts, put in place during past reforms to fight corruption. By the time cities can farm out a software project for competitive bidding, pick a winner, and issue a contract, a year or more might have gone by. They might not even need the app anymore. The winning bidder, most likely a freelancer or tiny independent software shop, might be swamped with other work or have gone out of business.

To augment cities' ability to source small software projects, Code for America acts as an intermediary. For a $180,000 annual fee, the organization provides each participating city with three fellows. After a month of training at the group's headquarters in San Francisco, followed by a month-long local immersion in their sponsor cities, the fellows return to California to work on projects directed by the sponsoring city. They receive a modest stipend of $35,000 plus health benefits for eleven months of service.

By 2012, when Pahlka's team put out the call for their third class of fellows, it was clear that Code for America was having a positive impact on the participating cities. When we spoke, she was traveling in Boston, and was buoyant about one of the projects completed there during the previous summer. Like many urban school districts that are trying to offer greater choice for parents, Boston has a maddeningly complex process for enrollment. Parents must study a twenty-eight-page pamphlet and manually map the radius of eligibility around their home (from one mile for elementary to up to two miles for high school), to find out which schools their children can apply to. As Nigel Jacob of the Office of New Urban Mechanics told me, "it unfolds into this strange map kind of a thing, very dense, very scary looking, very wordy."[18] The Boston school district had launched its own Web app a few years earlier, called "What Are My Schools?," that would spit out a simple list of schools based on a family's street address. It wasn't helping much, and by the summer of 2011, the *Boston Globe* turned the screws on Mayor Menino, running a series of scathing articles bashing the entire school assignment scheme.[19]

Schools were central to Menino's quality of life focus, and he made it clear to schools officials that something had to be done to address the problem quickly—"they got a very clear message from parents and the mayor," according to Jacob.[20] The Office of New Urban Mechanics stepped in, tasking one of its Code for America fellows to build a better tool for assessing school options. The result, a web app called Discover BPS, allowed parents to browse and sort a map of eligible schools that took into account all of the nagging and complex details of school selection, like a sibling's current school assignment.

Discover BPS was a big win for both Code for America and the Office of New Urban Mechanics' approach to innovation. The entire project, built by one fellow with assistance from two others, took less than four months from start to finish—an almost instant response, compared to the traditional way cities buy software. As Pahlka explains, the status quo "requires writing a specification, soliciting bids, qualifying contractors, and a lot of things that take a lot of time. Generally in city governments, a project like that will take about two years."

By comparison, in the private sector Web apps today can be put together in a week or less. The first version of Twitter was built in a month. The first version of Gmail was built in a day. From there, they evolve. Most Web start-ups now push out new code on a weekly basis, tweaking interfaces and debugging as they grow. As Pahlka describes, "Successful applications these days aren't completely spec'd out from the beginning and coded to that spec. They are built in a more agile process, more iteratively."

At least in Boston, Code for America is helping change the way people in city government think about creating new software for citizens. Discover BPS took "this complicated process, put it up on the Web, and made it work in such a way that the process is now simple, beautiful, and easy to use," Palhka says. But more importantly, "It proved that you could do it quickly, you could do it well, and you could do it relatively cheaply. If that's the case," she argues, "you start to create political will to question the traditional process." Top school officials were skeptical about Discover BPS in the beginning, according to Jacob. "Now they are 'big fans,'" he says.[21]

Pahlka's enthusiasm about the future is hard to resist. But when I first learned about Code for America, soon after its first call for fellows in 2010, I was skeptical. Gov 2.0 struck me as a vehicle for O'Reilly to promote his idea of "government as a platform," an ambitious but somewhat naïve proposal that seemed to want to dismantle the federal government and rebuild it with open-source software and open data. Let entrepreneurs, hackers, whoever step up and do the actual service delivery, the argument went, and make government a mere infrastructure provider.[22] For someone so publicly identified with the progressive left, it sounded like a techno-libertarian call to arms. "Government 2.0 is not a new kind of government; it is government stripped down to its core, rediscovered and reimagined as if for the first time," he wrote in a widely circulated essay. O'Reilly's tenacious support of open-source software made me suspect that Code for America was really just a ploy to box big software companies out of the government market. Governments all around the world were shifting to Linux, and pushing out Microsoft and IBM. Was O'Reilly plotting to bring the open-source fight back to the homeland?

Pahlka denies any ambition to go head-to-head with IBM, Oracle, or any of the other tech titans who have locked up the government software market. "We're too small—and going to stay small—to rewrite the technology landscape ourselves," she insists. To her, Code for America's work in cities is a demonstration, a disruption to business as usual. "We are not going to reconfigure IT systems at the city level from soup to nuts. We're not about a wholesale transition. We're about creating the stories and the examples other people can use to say, 'We can do it differently as well. How else can we apply this model? How can we get those results in that kind of time? What do we need to do to transform the way we work?'"[23]

But projects like the Discover BPS Web app are clearly direct replacements for software the city would previously have hired a contractor to build. And even as Pahlka claims to want to lead by inspiration alone, Code for America has used part of a $1.5 million grant received from Google in 2012 to fund what she describes as a "civic accelerator" in San Francisco. Its purpose? To incubate startups that disrupt the marketplace for government software. But by creating new companies that may end up competing directly with existing vendors for government contracts, the move could undermine Code for America's whole agenda around agile innovation. As Nigel Jacob at Boston's Office of New Urban Mechanics (who now sits on the Code for America board) put it, "Government as a platform . . . very much does sound like replacing one set of vendors with another set of vendors. The new set of vendors might be more lightweight, but eventually, once they have to start dealing with government contracting, they'll have to bulk up. At this point, they'll have to end up behaving a lot like some of these larger companies."[24]

In Pahlka's defense, a shake-up in the government tech business is long overdue, and the innovations the accelerator incubates might open up markets the technology giants have never even imagined. While "the accelerator's purpose is to create businesses that will disrupt the current government technology ecosystem," she told me, "the more disruptive ones will actually just go direct to citizens."[25]

Code for America is an exemplar of what I call "computational leadership networks," which are national and international organizations that go beyond just sharing stories and case studies of smart-city innovations. A thicket of international intercity organizations already exist for that purpose, issuing endless streams of reports and organizing costly, often pointless junkets.[26] Instead, these new networks help cities share real resources—actual working code, models, and data. The intensity of this exchange is evident in Code for America's 2011 stats: 21 civic apps produced, 12,828 code commits (a measure of programmer productivity), 390 civic leaders engaged, 546 code community members registered.[27] It's a fundamental challenge to the big companies who have traditionally extracted a hefty profit for the service of porting solutions from one city to another.

The other challenge, which we'll examine more deeply in the next section, is the parochialism in software procurement that's found in city governments everywhere. City-led efforts to innovate new digital services are usually tied to some effort to spearhead development of a local technology industry. The technology people in city government may understand the value of simply repurposing a system from another city or an existing company. But the economic-development officials want to see government contracts spent locally. As a result, many city-funded technology projects end up reinventing the wheel. For Pahlka the challenge is to "Get cities to get over this notion that 'this has to be about our city.' You need to be a leader in *cooperating* with other cities." Sometimes, too much situated software is a bad thing.

Code for America's biggest challenge to growth, just as we saw with IBM in chapter 2, is scaling its business model and technology down. It's a big, rich city model that requires a well-funded tech staff and infrastructure to support its fellows. How will it work in the thousands of communities of ten thousand or fifty thousand or a hundred thousand whose civil servants are often supported by a single IT person? Pahlka's solution is to use the rest of Google's money to launch the Code for America Brigade, an online community to connect individuals who want to deploy Code for America apps in their communities and contribute back to the common code base. "We're not going to fix government until we fix citizenship," she says. In her future, knowing how to code will be an important skill for civic improvement.

Not Invented Here

Sascha Haselmayer is livid. "That solution I sent you for the blind is just astounding," he raves. "Just in New York, it would allow 380,000 people to navigate completely independently through the city for the first time in human history."[28]

It was a pretty remarkable gadget. Invented by Swedish firm Astando, e-Adept was financed in part by the city of Stockholm in its quest to become, according to the city's website, "the most accessible capital in the world."[29] Using an exquisitely detailed digital map of the city's terrain, the GPS-enabled headset talks to the user, calling out obstacles and safe paths. "It has had a huge impact—empowering those people to find jobs, releasing their relatives, and reducing demand on social services," Haselmayer says. He claims that for just $500,000 in annual operating costs, the system is generating $20 million a year in direct economic benefits for Stockholm.

Haselmayer is the founder of another start-up that's cross-fertilizing smart-city innovation from its base in Barcelona, Living Labs Global. Earlier that year, Haselmayer had pitched e-Adept to city officials in New York. But their response was the same as many other cities. "If you put something like that on the table of any CIO in any city," he laments, referring to a relatively new high-level executive position being created in many cities, the Chief Information Officer, "they will say it doesn't fit into their architecture." What they mean is that it's not a priority, not worth the hassle of making it work with their existing systems. Haselmeyer sighs. "Do you think it will fit into the lives of 380,000 people in your city? To get up in the morning and go to work?" I can tell from his tone he didn't close the deal.

As I speak with him by Skype from his office in Barcelona, Haselmayer paints a convincing picture of situated software gone wrong. "Look at Germany. You have twenty-four cities which each have their own mobile app for parking. Every city backs its own local service provider thinking that they're helping the next Google to emerge. They reinvent the wheel and dress it up as a big local innovation program." Across Europe, he has discovered fifty-six cities that have built their own bad variations of the same service. And not only are citizens stuck with subpar apps, they need to use a different one every time they drive to the next town.

Meanwhile, the Estonian firm that invented mobile parking in the first place has struggled to grow for over a decade. After the success of its ParkNOW! service, which launched in the Baltic nation's capital of Tallinn in 2000, NOW! Innovations had pitched "a thousand cities around the world," Haselmayer explains. "In every country they had to hire a local representative to actually present the project for them. They spent almost $10 million dollars on marketing." Despite being the first business to enter a market that Haselmayer estimates could be as big as $65 billion globally, the company had grown at a snail's pace.

This ineffective duplication of smart-city technologies is a global problem. "Every city orders an innovation project to invent something without actually seeing what has been done before elsewhere," Haselmayer explained. "You can see that in the worst-managed cities and in the best-managed cities . . . where they spend hundreds of millions reinventing everything from scratch. Absolutely everything." In *Connected Cities*, a book he coauthored with his Living Labs Global cofounders, he estimated that it was costing European cities tens of millions of dollars each year in duplicated efforts.[29] Of smart-city entrepreneurs, he tells me, "We're killing them one after another, and then they end up doing ringtones for BlackBerrys, because they know how to make money with that."

But why weren't the people who invented mobile parking able to succeed? Haselmayer has catalogued hundreds of start-ups and entrepreneurs around the world with cutting-edge smart technologies, and what he discovered is that they all suffer from the same visibility problem and a "not invented here" attitude among their potential customers abroad. "How can a city trust someone approaching them and saying 'I've invented mobile parking. You should help me make it happen.'" What smart-city start-ups needed was a cost-effective way to market themselves outside their hometowns and compete with the big technology giants.

Haselmayer set out to design a fix. In 2010 he drafted a handful of cities to issue challenges, and invited his network of start-ups to show how their technology could address them. The Living Labs Global Awards, which entered its fourth year in 2013, are selected by a jury convened by each city. After the contest, cities can engage the winner to implement the solution, or write the affair off as a brainstorming exercise. The award was designed to "give these companies visibility, help them to get an opening internationally." When we spoke in late 2011, there were signs that the model was working—he reported that pilots based on winning projects in 2011 were up and running in Chicago, Taipei, and Lagos.

A few months after our conversation over Skype, I met up with Haselmayer in Barcelona. As we wove our way across the old city, ducking in and out of medieval plazas, Haselmayer beamed as he explained the latest thrust in his campaign to promote smart-city start-ups, a new website called CityMart. He recited his pitch: "It's a platform that provides cities with market intelligence about what kind of solutions are being developed, and where they are working." "You mean it's an Amazon for smart cities?" I asked. "Exactly!" he said, grinning.[30]

Since we had last spoken, I'd thought often about Germany's parking app fiasco. As much as I believed that the organic approach to smart-city innovation was better in the long run, Haselmayer's story had raised serious concerns about the wisdom of building smart city technology locally. I'd embraced the notion of civic laboratories as factories for situated software, of quirky local apps, and infrastructures that put a unique local spin on technology. Twenty years of studying cities told me that building small, local, and human-scale was always better. But his research showed that most cities didn't actually have the capacity to create good apps. Perhaps I hadn't appreciated how hard it is for good technology to spread and take root where it was needed.

Haselmayer's slant on situated software cuts across geography instead of rooting itself in individual places. "There are 557,000 local governments in the world," he told me, "and they just cannot be all different. I'm trained as an urbanist so it's not that I reject the idea that every place is unique." Instead, he saw an opportunity to target "micromarkets," as he calls them, that don't present a huge opportunity in one city but are potentially enormous if you can aggregate them globally. "Blind people everywhere have the same problems," he says, recalling Stockholm's e-Adept system.

At a sidewalk cafe, I prodded Haselmayer, picking up that thread again. He pulled up CityMart on his iPad. The website lets you travel virtually around the world and see how other cities have solved similar problems. A manager in a city's parking authority, having identified the need for a mobile payment system, might browse through CityMart's dozens of showcases from different companies from around the Web. "There's this romantic public-sector view that you do a study trip, you like the ideas, and then you go home and do the same thing," he said. This was the main way that ideas spread before the Web was invented. "This is not so cities can exchange best practices," he told me, pointing at CityMart on the iPad. There are already lots of international organizations that ply a rich trade in urban-planning case studies. "It is so cities can exchange contractors." His goal is for the site to eventually house five thousand companies offering technology solutions for every urban problem under the sun.

CityMart should help speed the spread of smart-city technologies. Potentially, it could transform the industry, making it less top-heavy and less dominated by global firms like IBM, Cisco, and Siemens. It's less clear whether it will do much about the "not invented here" problem. If it does truly create a new global trade in smart city solutions, local officials may be under more pressure than ever to make sure their dollars go to local firms that could themselves use CityMart for a real shot at larger success.

My phone buzzed with directions to my next appointment. That evening I was using Barcelona's cafés and bars as a kind of virtual conference center, all coordinated through my Foursquare social graph. Haselmayer offered his cynical view of the smart-cities industry, which had gathered in Barcelona for one of its biggest global trade shows to date. "The debate on smart cities has become all about [technical] architecture, where IBM says a smart city is nothing else but a corporation, and you need a good kind of architecture and then everything happens.

That is an unrealistic view of how a city works, and it's a monolithic approach. They are saying that you don't prioritize by deploying services, you prioritize by building yet another municipal industrial-scale infrastructure. Once you've got this, then everything is possible."

He pushed a copy of his book across the table. I flipped through the evidence for his case, a painstaking accounting of billions of euros of public monies wasted on useless apps. "Where are the services that can change our lives?" the jacket asks. For citizens of the world's emerging smart cities, it's the right question. But as Bill Clinton has said, "Nearly every problem has been solved by someone somewhere. The challenge of the twenty-first century is to find out what works and scale it up."[31]

The Long Hack

The Summer of Love was a rejection of the material abundance of America's new middle class. The hippies of the Haight questioned the very foundations of capitalist society—property, marriage, and even government itself. They abhorred industrial systems of production, and tried to re-create local alternatives. Similarly, in an ideal world, we'd craft vernacular technologies to meet the unique needs of every smart city using only the materials at hand locally. We'd slow down and open up the design process, to ensure maximum participation by the people who will live with them for a century to come. This is what Summer of Smart was after.

Unfortunately, we don't have the time to tailor a bespoke set of smart technologies for every city. Many are growing much too rapidly for that organic process to play out, while others slip speedily into decline. Technology is evolving even more rapidly, creating new tools to address these problems and rapidly making the old solutions obsolete. If we are to realize the opportunity smart technologies present, global industry has to play a role. Grassroots movements can be innovative and powerful, but just as often they are slow, factionalized, disorganized, and disorderly. A half-decade after Washington,

DC, opened up the first municipal open-data store, according to my calculations, less than 6 percent of the US population lived in a place that had one.[32] And the diffusion of new ideas to small cities and those in the global south is not happening fast enough. Smart is still mostly a big-city phenomenon.

This frustratingly slow pace of progress is fueling a growing urge to create standards for smart cities. How will a building's systems talk to each other? How can my phone ask a bus where it is going? Companies like Living PlanIT, based in London, talk openly about their ambition to develop an "urban operating system." But the engineering challenge of making all the varied pieces of technology in a city work together, as enormous as they are, is just the first baby step in rationalizing the design of the smart city. Already a consortium of cities led by Barcelona (with a strong Cisco presence) convened in 2012 to start work on a "City Protocol," which aims to create not only technical standards but also a common language to describe "the anatomy, the functions and the metabolism of a city" and performance indicators to measure and benchmark them.[33]

A common new starting point for building smart cities will speed the diffusion of good ideas and technology. But in the rush to set standards, we should heed the lessons of those early struggles over the Internet's DNA that we explored in chapter 3. For if the lessons of these civic laboratories and the situated software they generate tell us anything, it is to be careful how much structure we impose from the top down. The Internet's development shows how the combinatorial approach to innovation, though by nature incremental, can add up to big breakthroughs that quickly scale planet-wide. The endless variety of pilots, prototypes, and experiments popping up across the globe demonstrates that this style of combinatorial innovation is alive and well in the realm of smart cities. Every day, tinkerers around the world are showing that smart technologies are a very different beast than mere urban utilities. They are complex assemblages crafted to solve the everyday needs of small groups of people. With luck, just like the Web, over time these small, localized advances will add up to big positive changes in how we all live and work. Perhaps we should hold our options open a bit longer and resist the urge to standardize too much.

In 2010, Geoffrey West, the physicist who studies cities, remarked at a gathering of urban scholars in New York that if we don't have a science of cities, "then all cities need to be dealt with individually."[34] But for designers, dealing with cities individually is the only proper approach. This growing tension between expedient deployment and careful design in smart cities isn't going away. Every city is its own sticky knot of people, places, and policies. Even if every smart city was crafted from a common template, it will need to be customized to get the right fit with the existing city. Every city will have to strike a balance based on its patience, its financial resources, and its capacity to innovate locally.

Clearly, this is going to take time. We should settle in for the long hack.

Like the Internet, this planet of civic laboratories is destined to become more than the sum of its parts as ideas circulate within and between cities. But how the balance between local innovation and cross-fertilization plays out is still unclear. If Shirky's forecast is right, and the future is filled by millions of apps that nail the needs of small groups, the combinatorial approach will dominate. If industry is right, success will stem from the spread of a handful of core breakthroughs and standards.

Somewhere in the middle is the more realistic future, a weblike global network of smart cities, swapping ideas, tools, and data in real time. But to make that happen, we'll need to get better at extracting the repurposable improvements from situated software that can be cross-fertilized elsewhere. We'll need more ways to share them between cities, faster ways to graft them onto new places, and at least some universal standards to make the process as cheap as possible. And we'll have to manage all of this without preempting too many of the design decisions that should be made locally.

Overstandardization could weed out too much of the competitive urge that has driven creativity and innovation throughout history—

there's a long precedent for this planet of civic laboratories. In 1948 the great British philosopher Bertrand Russell explained how arts had flourished before industrialization through intercity competition. "The inferiority of our age," in integrating the arts into everyone's daily lives, he said, "is an inevitable result of the fact that society is centralised and organised to such a degree that individual initiative is reduced to a minimum. Where art has flourished in the past it has flourished as a rule amongst rival small communities, such as the Greek City States, the little Principalities of the Italian Renaissance, and the petty Courts of German eighteenth-century rulers." Russell yearned for the dynamic that's at play in smart cities today. "It would be a good thing if cities could develop an artistic pride leading them to mutual rivalry, and if each had its own school of music and painting, not without a vigorous contempt for the school of the next city. . . . I think that this problem of giving importance to localities will have to be tackled if human life is not to become increasingly drab and monotonous."[35]

Our standards must be set with care, for unlike consumers, cities won't be able to just throw their legacy technologies away when they become obsolete. The consequences of decisions made today will be with us for years and even decades to come. As Eran Ben-Joseph, a scholar of urban design at MIT, has written, standards for subdividing land, laying utilities, and configuring streets and sidewalks that were promulgated a century ago in the name of progress now constrain us from addressing new problems. "Originating in the desire to improve conditions in urban areas in the late nineteenth and early twentieth centuries, standards became the essential tool for solving the problems of health, safety and morality. . . . Because so much has been built according to these dictates, the accumulated rules now have the force of universal acceptance—standards have become the definers, delineators, and promoters of places, regardless of variations in landform, natural systems, and human culture."[36]

Should we rush too quickly to lock in the design of our smart cities and the technology that powers them, we may miss the last and greatest chance to recapture the elaborate diversity that makes them special.

9

Buggy, Brittle, and Bugged

Calafia Café in Palo Alto is one of the smartest eateries in the world. With Google's former executive chef Charlie Ayers at the helm, the food here isn't just for sustenance. This is California—eating is also a path to self-improvement. Each dish is carefully crafted with ingredients that not only keep you slim, but make you smarter and more energized too. A half-dozen venture capitalists pick at their dandelion salads. A sleepy suburb at night, by day Palo Alto becomes the beating heart of Silicon Valley, the monied epicenter of the greatest gathering of scientific and engineering talent in the history of human civilization. To the west, across the street, lies Stanford University. The Googleplex sprawls a few miles to the east. In the surrounding region, some half-million engineers live and work. A tech tycoon or two wouldn't be out of place here. Steve Jobs was a regular.

Excusing myself to the men's room, however, I discover that Calafia Café has a major technology problem. Despite the pedigree of its clientele, the smart toilet doesn't work. As I stare hopefully at the stainless steel throne, a red light peering out from the small black plastic box that contains the bowl's "brains" blinks at me fruitlessly. Just above, a sign directs an escape path. "If sensor does not work," it reads, "use manual flush button." And so I bail out, sidestepping fifty years of progress in computer science and industrial engineering in the blink of an eye.

Back at my table, I try to reverse-engineer the model of human-waste production encoded in the toilet's CPU. I imagine a lab somewhere in Japan. Technicians in white lab coats wield stopwatches as they methodically clock an army of immodest volunteers seated upon row after row of smart johns. The complexity of the problem becomes clear. Is it supposed to flush as soon as you stand up? Or when you turn around? Or pause for a fixed amount of time? But how long? Can it tell if you need another flush? It's not quite as challenging an engineering task as putting a man on the moon, or calculating driving directions to the airport. Somehow, though, that stuff works every time.

My bewilderment quickly yields to a growing sense of dread. How is it that even in the heart of Silicon Valley it's completely acceptable for smart technology to be buggy, erratic, or totally dysfunctional? Someone probably just cured cancer in the biotechnology lab across the street and is here celebrating over lunch. Yet that same genius will press the manual flush button just as I did, and never think twice about how consistently this new world of smart technology is letting us down. We are weaving these technologies into our homes, our communities, even our very bodies—but even experts have become disturbingly complacent about their shortcomings. The rest of us rarely question them at all.

I know I should stop worrying, and learn to love the smart john. But what if it's a harbinger of bigger problems? What if the seeds of smart cities' own destruction are already built into their DNA? Up to this point, I've argued that smart cities are a solution to the challenges of twenty-first-century urbanization. I've told you that despite potential pitfalls, the benefits outweigh the risks, especially if we are aggressive about confronting the unintended consequences of our choices. But in reality we've only scratched the surface.

What if the smart cities of the future are buggy, brittle, and bugged? What are we getting ourselves into?

Buggy

A few weeks later, I found myself wandering around the MIT campus in Cambridge, Massachusetts, with nary a thought about uncooperative toilets in mind. Strolling west from Kenmore Square, a few minutes later I came across the new home of the Broad Institute, a monolith of glass and steel that houses a billion-dollar center for research in genomic medicine. The street wall was tricked out with an enormous array of displays showing in real time the endless sequences of DNA base pairs being mapped by the machinery upstairs.

And then, out of the corner of my eye, I saw it. The Blue Screen of Death, as the alert displayed by Microsoft Windows following an operating-system crash is colloquially known. Forlorn, I looked through the glass at the lone panel. Instead of the stream of genetic discoveries, a meaningless string of hexadecimals stared back, indicating precisely where, deep in the core of some CPU, a lone miscomputation had occurred. Just where I had hoped to find historic fusion of human and machine intelligence, I'd found yet another bug.

The term "bug," derived from the old Welsh *bwg* (pronounced "boog"), has long been used as slang for insects. But appropriation of the term to describe technical failings dates to the dawn of the telecommunications age. The first telegraphs invented in the 1840s used two wires, one to send and one to receive. In the 1870s, duplex telegraphs were developed, permitting messages to be sent simultaneously in both directions over a single wire. But sometimes stray signals would come down the line, which were said to be "bugs" or "buggy."[1] Thomas Edison himself used the expression in an 1878 letter to Puskás Tivadar, the Hungarian inventor who came up with the idea of a telephone exchange that allowed individual lines to be connected into a network for the first time.[2] According to an early history of Edison's own quadruplex, an improved telegraph that could send two signals in each direction, by 1890 the word had become common industry parlance.[3]

The first documented computer bug, however, was an actual insect. In September 1947, Navy researchers working with professors at Harvard University were running the Mark II Aiken Relay Calculator through its paces when it suddenly began to miscalculate. Tearing open the primitive electromechanical computer, they found a moth trapped between one of its relays. On a website maintained by Navy historians, you can still see a photograph of the page from the lab notebook where someone carefully taped the moth down, methodically adding an annotation: "First actual case of bug being found."[4] As legend has it, that person was Grace Hopper, a programmer who would go on to become an important leader in computer science. (Hopper's biographer, however, disputes this was the first time "bug" was used to describe a malfunction in the early development of computers, arguing "it was clear the term was already in use.")[5]

Since that day, bugs have become endemic in our digital world, the result of the enormous complexity and ruthless pace of modern engineering. But how will we experience bugs in the smart city? They could be as isolated as that faulty toilet or a crashed public screen. In 2007 a Washington Metro rail car caught fire after a power surge went unnoticed by buggy software designed to detect it.[6] Temporarily downgrading back to the older, more reliable code took just twenty minutes per car while engineers methodically began testing and debugging.

But some bugs in city-scale systems will ripple across networks with potentially catastrophic consequences. A year before the DC Metro fire, a bug in the control software of San Francisco's Bay Area Rapid Transit (BART) system forced a systemwide shutdown not just once, but three times over a seventy-two-hour period. More disconcerting is the fact that initial attempts to fix the faulty code actually made things worse. As an official investigation later found, "BART staff began immediately working to configure a backup system that would enable a faster recovery from any future software failure." But two days after the first failure, "work on that backup system inadvertently contributed to the failure of a piece of hardware that, in turn, created the longest delay."[7] Thankfully, no one was injured by these subway shutdowns, but their economic impact was likely enormous—the economic toll of the two-and-a-half-day shutdown of New York's subways during a 2005 strike was estimated at $1 billion.[8]

The troubles of automation in transit systems are a precursor to the kinds of problems we're likely to see as we buy into smart cities. As disconcerting as today's failures are, however, they are actually a benchmark for reliability. Current smart systems are painstakingly designed and extensively tested. They have multiple layers of fail-safes. With the urgency of urban problems increasing and the resources and will to deal with them in doubt, in the future many smart technologies will be thrown together under tight schedules and even tighter budgets. They will struggle to match this gold standard of reliability, with only a few short-lived, sporadic glitches each year.

The sheer size of city-scale smart systems comes with its own set of problems. Cities and their infrastructure are already the most complex structures humankind has ever created. Interweaving them with equally complex information processing can only multiply the opportunities for bugs and unanticipated interactions. As Kenneth Duda, a high-performance networking expert told the *New York Times*, "the great enemy is complexity, measured in lines of code, or interactions."[9] Ellen Ullman, a writer and former software developer, argues, "it is impossible to fully test any computer system. To think otherwise is to misunderstand what constitutes such a system. It is not a single body of code created entirely by one company. Rather, it is a collection of 'modules' plugged into one another. . . . The resulting system is a tangle of black boxes wired together that communicate through dimly explained 'interfaces.' A programmer on one side of an interface can only hope that the programmer on the other side has gotten it right."[10]

In his landmark 1984 study of technological disasters, *Normal Accidents*, sociologist Charles Perrow argued that in highly complex systems with many tightly linked elements, accidents are inevitable. What's worse is that traditional approaches to reducing risk, such as warnings and alerts (or the installation of the backup recovery system in the BART incident), may actually introduce more complexity into systems and thereby increase risks. The Chernobyl nuclear disaster, for instance, was caused by an irreversible chain of events triggered during tests of a new reactor safety system. Perrow's conclusion: "Most high-risk systems have some special characteristics, beyond their toxic or explosive or genetic dangers, that make accidents in them inevitable, even 'normal.' "[11]

Normal accidents will be ever-present in smart cities. Just as the rapid pace of urbanization has revealed shoddy construction practices, most notably in China's notorious "tofu buildings," hastily put together smart cities will have technological flaws created by designers' and builders' shortcuts. These hasty hacks threaten to make earlier design shortcuts like the Y2K bug seem small in comparison. Stemming from a trick commonly used to save memory in the early days of computing, by recording dates using only the last two digits of the year, Y2K was the biggest bug in history, prompting a worldwide effort to rewrite millions of lines of code in the late 1990s. Over the decades, there were plenty of opportunities to undo Y2K, but thousands of organizations chose to postpone the fix, which ended up costing over $300 billion worldwide when they finally got around to it.[12] Bugs in the smart city will be more insidious, living inside lots of critical, interconnected systems. Sometimes there may be no way to anticipate the interdependencies. Who could have foreseen the massive traffic jam caused on US Interstate 80 when a bug in the system used to manage juror pools by Placer County, California, erroneously summoned twelve hundred people to report for duty on the same day in 2012?[13]

The pervasiveness of bugs in smart cities is disconcerting. We don't yet have a clear grasp of where the biggest risks lie, when and how they will cause systems to fail, or what the chain-reaction consequences will be. Who is responsible when a smart city crashes? And how will citizens help debug the city? Today, we routinely send anonymous bug reports to software companies when our desktop crashes. Is this a model that's portable to the world of embedded and ubiquitous computing?

Counterintuitively, buggy smart cities might strengthen and increase pressure for democracy. Wade Roush, who studied the way citizens respond to large-scale technological disasters like blackouts and nuclear accidents, concluded that "control breakdowns in large technological systems have educated and radicalized many lay citizens, enabling them to challenge both existing technological plans and the expertise and authority of the people who carry them out." This public reaction to disasters of our own making, he argues, has spurred the development of "a new cultural undercurrent of 'technological citizenship' characterized by greater knowledge of, and skepticism toward, the complex systems that permeate modern societies."[14] If the first generation of smart cities does truly prove fatally flawed, from their ashes may grow the seeds of more resilient, democratic designs.

In a smart city filled with bugs, will our new heroes be the adventurous few who can dive into the ductwork and flush them out? Leaving the Broad Institute's Blue Screen of Death behind, I headed back in the rain to my hotel, reminded of *Brazil*, the 1985 film by Monty Python troupe member Terry Gilliam, which foretold an autocratic smart city gone haywire. Arriving at my room, I opened my laptop and started up a Netflix stream of the film. As the scene opens, the protagonist, Sam Lowry, squats sweating by an open refrigerator. Suddenly the phone rings, and Harry Tuttle, played by Robert De Niro, enters. "Are you from Central Services?" asks Lowry, referring to the uncaring bureaucracy that runs the city's infrastructure. "They're a little overworked these days," Tuttle replies. "Luckily I intercepted your call." Tuttle is a guerrilla repairman, a smart-city hacker valiantly trying to keep residents' basic utilities up and running. "This whole system of yours could be on fire, and I couldn't even turn on a kitchen tap without filling out a twenty-seven-B-stroke-six."

Let's hope that's just a story. Some days, it doesn't feel so far-fetched.

Brittle

Creation myths rely on faith as much as fact. The Internet's is no different. Today, netizens everywhere believe that the Internet began as a military effort to design a communications network that could survive a nuclear attack.

The fable begins in the early 1960s with the publication of "On Distributed Communications" by Paul Baran, a researcher at the RAND think tank. At the time, Baran had been tasked with developing a scheme for an indestructible telecommunications network for the US Air Force. Cold War planners feared that the hub-and-spoke structure of the telephone system was vulnerable to a preemptive Soviet first strike. Without a working communications network, the United States would not be able to coordinate a counterattack, and the strategic balance of "mutually assured destruction" between the superpowers would be upset. What Baran proposed, according to Harvard University science historian Peter Galison, "was a plan to remove, completely, critical nodes from the telephone system."[15] In "On Distributed Communications" and a series of pamphlets that followed, he demonstrated mathematically how a less centralized latticework of network hubs, interconnected by redundant links, could sustain heavy damage without becoming split into isolated sections.[16] The idea was picked up by the Pentagon's Advanced Research Projects Agency (ARPA), a group set up to fast-track R&D after the embarrassment of the Soviet space program's Sputnik launch in 1957. ARPANET, the Internet's predecessor, was rolled out in the early 1970s.

So legend has it.

The real story is more prosaic. There were indeed real concerns about the survivability of military communications networks. But RAND was just one of several research groups that were broadly rethinking communications networks at the time—parallel efforts on distributed communications were being led by Lawrence Roberts at MIT and Donald Davies and Roger Scantlebury at the United Kingdom's National Physical Laboratory. Each of the three efforts remained unaware of each other until a 1967 conference organized by the Association for Computing Machinery in Gatlinburg, Tennessee, where Roberts met Scantlebury, who by then had learned of Baran's earlier work.[17] And ARPANET wasn't a military command network for America's nuclear arsenal, or any arsenal for that matter. It wasn't even classified. It was a research network. As Robert Taylor, who oversaw the ARPANET project for the Pentagon, explained in 2004 in a widely forwarded e-mail, "The creation of the ARPAnet was not motivated by considerations of war. The ARPAnet was created to enable folks with common interests to connect to one another through interactive computing even when widely separated by geography."[18]

We also like to think that the Internet is still widely distributed as Baran envisioned, when in fact it's perhaps the most centralized communications network ever built. In the beginning, ARPANET did indeed hew closely to that distributed ideal. A 1977 map of the growing network shows at least four redundant transcontinental routes, run over phone lines leased from AT&T, linking up the major computing clusters in Boston, Washington, Silicon Valley, and Los Angeles. Metropolitan loops created redundancy within those regions as well.[19] If the link to your neighbor went down, you could still reach them by sending packets around in the other direction. This approach is still commonly used today.

By 1987, the Pentagon was ready to pull the plug on what it had always considered an experiment. But the research community was hooked, so plans were made to hand over control to the National Science Foundation, which merged the civilian portion of the ARPANET with its own research network, NSFNET, launched a year earlier. In July 1988, NSFNET turned on a new national backbone network that dropped the redundant and distributed grid of ARPANET in favor of a more efficient and economical hub-and-spoke arrangement.[20] Much like the air-transportation network today, consortia of universities pooled their resources to deploy their own regional feeder networks (often with significant NSF funding), which linked up into the backbone at several hubs scattered strategically around the country.

Just seven years later, in April 1995, the National Science Foundation handed over management of the backbone to the private sector. The move would lead to even greater centralization, by designating just four major interconnection points through which bits would flow across the country. Located outside San Francisco, Washington, Philadelphia, and Chicago, these hubs were the center not just of America's Internet, but the world's. At the time, an e-mail from Europe to Asia would almost certainly transit through Virginia and California. Since then, things have centralized even more. One of those hubs, in Ashburn, Virginia, is home to what is arguably the world's largest concentration of data centers, some forty buildings boasting the collective footprint of twenty-two Walmart Supercenters.[21] Elsewhere, Internet infrastructure has coalesced around preexisting hubs of commerce. Today, you could knock out a handful of buildings in Manhattan where the world's big network providers connect to each other—60 Hudson Street, 111 Eighth Avenue, 25 Broadway—and cut off a good chunk of transatlantic Internet capacity. (Fiber isn't the first technology to link 25 Broadway to Europe. The elegant 1921 edifice served as headquarters and main ticket office for the great ocean-crossing steamships of the Cunard Line until the 1960s.)

Despite the existence of many chokepoints, the Internet's nuke-proof design creation myth has only been strengthened by the fact that the few times it has actually been bombed, it has proven surprisingly resilient. During the spring 1999 aerial bombardment of Serbia by NATO, which explicitly targeted telecommunications facilities along with the power grid, many of the country's Internet Protocol networks were able to stay connected to the outside world.[22] And the Internet survived 9/11 largely unscathed. Some 3 million telephone lines were knocked out in lower Manhattan alone—a grid the size of Switzerland's—from damage to a single phone-company building near the World Trade Center. Broadcast radio and TV stations were crippled by the destruction of the north tower, whose rooftop bristled with antennas of every size, shape, and purpose. Panic-dialing across the nation brought the phone system to a standstill.[23] But the Internet hardly blinked.

But while the Internet manages to maintain its messy integrity, the infrastructure of smart cities is far more brittle. As we layer ever more fragile networks and single points of failure on top of the Internet's still-resilient core, major disruptions in service are likely to be common. And with an increasing array of critical economic, social, and government services running over these channels, the risks are compounded.

The greatest cause for concern is our growing dependence on untethered networks, which puts us at the mercy of a fragile last wireless hop between our devices and the tower. Cellular networks have none of the resilience of the Internet. They are the fainting ladies of the network world—when the heat is on, they're the first to go down and make the biggest fuss as they do so.

Cellular networks fail in all kinds of ugly ways during crises: damage to towers (fifteen were destroyed around the World Trade Center on 9/11 alone), destruction of the "backhaul" fiber-optic line that links the tower into the grid (many more), and power loss (most towers have just four hours of battery backup). In 2012, flooding caused by Hurricane Sandy cut backhaul to over two thousand cell sites in eight counties in and around New York City and its upstate suburbs (not including New Jersey and Connecticut), and power to nearly fifteen hundred others.[24] Hurricane Katrina downed over a thousand cell towers in Louisiana and Mississippi in August 2005, severely hindering relief efforts because the public phone network was the only common radio system among many responding government agencies. In the areas of Japan north of Tokyo annihilated by the 2011 tsunami, the widespread destruction of mobile-phone towers literally rolled the clock back on history, forcing people to resort to radios, newspapers, and even human messengers to communicate. "When cellphones went down, there was paralysis and panic," the head of emergency communications in the city of Miyako told the *New York Times*.[25]

The biggest threat to cellular networks in cities, however, is population density. Because wireless carriers try to maximize the profit-making potential of their expensive spectrum licenses, they typically only build out enough infrastructure to connect a fraction of their customers in a given place at the same time. "Oversubscribing," as this carefully calibrated scheme is known in the business, works fine under normal conditions, when even the heaviest users rarely chat for more than a few hours a day. But during a disaster, when everyone starts to panic, call volumes surge and the capacity is quickly exhausted. On the morning of September 11, for instance, fewer than one in twenty mobile calls were connected in New York City.[26] A decade later, little has changed. During a scary but not very destructive earthquake on the US East Coast in the summer of 2011, cell networks were again overwhelmed. Yet media reports barely noted it. Cellular outages during crises have become so commonplace in modern urban life that we no longer question why they happen or how the problem can be fixed.

Disruptions in public cloud-computing infrastructure highlight the vulnerabilities of dependence on network apps. Amazon Web Services, the eight-hundred-pound gorilla of public clouds that powers thousands of popular websites, experienced a major disruption in April 2011, lasting three days. According to a detailed report on the incident posted to the company's website, the outage appears to have been a normal accident, to use Perrow's term. A botched configuration change in the data center's internal network, which had been intended to upgrade its capacity, shunted the entire facility's traffic onto a lower-capacity backup network. Under the severe stress, "a previously unencountered bug" reared its head, preventing operators from restoring the system without risk of data loss.[27] Later, in July 2012, a massive electrical storm cut power to the company's Ashburn data center, shutting down two of the most popular Internet services—Netflix and Instagram.[28] "Amazon Cloud Hit By Real Cloud," quipped a *PC World* headline.[29]

The cloud is far less reliable than most of us realize, and its fallibility may be starting to take a real economic toll. Google, which prides itself on high-quality data-center engineering, suffered a half-dozen outages in 2008 lasting up to thirty hours.[30] Amazon promises its cloud customers 99.5 percent annual uptime, while Google pledges 99.9 percent for its premium apps service. That sounds impressive until you realize that even after years of increasing outages, even in the most blackout-prone region (the Northeast), the much-maligned American electric power industry averages 99.96 percent uptime.[31] Yet even that tiny gap between reality and perfection carries a huge cost. According to Massoud Amin of the University of Minnesota, power outages and power quality disturbances cost the US economy between $80 billion and $188 billion a year.[32] A back-of-the-envelope calculation published by International Working Group on Cloud Computing Resiliency tagged the economic cost of cloud outages between 2007 and mid-2012 at just $70 million (not including the July 2012 Amazon outage).[33] But as more and more of the vital functions of smart cities migrate to a handful of big, vulnerable data centers, this number is sure to swell in coming years.

Cloud-computing outages could turn smart cities into zombies. Biometric authentication, for instance, which senses our unique physical characteristics to identify individuals, will increasingly determine our rights and privileges as we move through the city—granting physical access to buildings and rooms, personalizing environments, and enabling digital services and content. But biometric authentication is a complex task that will demand access to remote data and computation. The keyless entry system at your office might send a scan of your retina to a remote data center to match against your personnel record before admitting you. Continuous authentication, a technique that uses always-on biometrics—your appearance, gestures, or typing style—will constantly verify your identity, potentially eliminating the need for passwords.[34] Such systems will rely heavily on cloud computing, and will break down when it does. It's one thing for your e-mail to go down for a few hours, but it's another thing when everyone in your neighborhood gets locked out of their homes.

Another "cloud" literally floating in the sky above us, the Global Positioning System satellite network, is perhaps the greatest single point of failure for smart cities. Without it, many of the things on the Internet will struggle to ascertain where they are. America's rivals have long worried about their dependence on the network of twenty-four satellites owned by the US Defense Department. But now even America's closest allies worry that GPS might be cut off not by military fiat but by neglect. With a much-needed modernization program for the decades-old system way behind schedule, in 2009 the Government Accountability Office lambasted the Air Force for delays and cost overruns that threatened to interrupt service.[35] And the stakes of a GPS outage are rising fast, as navigational intelligence permeates the industrial and consumer economy. In 2011 the United Kingdom's Royal Academy of Engineering concluded that "a surprising number of different systems already have GPS as a shared dependency, so a failure of the GPS signal could cause the simultaneous failure of many services that are probably expected to be independent of each other."[36] For instance, GPS is extensively used for tracking suspected criminals and land surveying. Disruptions in GPS service would require rapidly reintroducing older methods and technologies for these tasks. While alternatives such as Russia's GLONASS already exist, and the European Union's Galileo and China's Compass systems will provide more alternatives in the future, the GPS seems likely to spawn its own nasty collection of normal accidents. "No-one has a complete picture," concluded Martyn Thomas, the lead investigator on the UK study, "of the many ways in which we have become dependent on weak signals 12,000 miles above us."[37]

Centralization of smart-city infrastructure is risky, but decentralization doesn't always increase resilience. Uncoordinated management can create its own brittle structures, such as the Internet's "bufferbloat" problem. Buffering, which serves as a kind of transmission gearbox to sync fast-flowing and congested parts of the Internet, is a key tool to smoothing out surges of data and reducing errors. But in 2010 Jim Gettys, a veteran Internet engineer, noticed that manufacturers of network devices had taken advantage of rapidly falling memory prices to beef up buffers far beyond what the Internet's original congestion-management scheme was designed for. "Manufacturers have reflexively acted to prevent any and all packet loss and, by doing so, have inadvertently defeated a critical TCP congestion-detection mechanism," concluded the editors of *ACM Queue*, a leading computer networking journal, referring to the Internet's traffic cop, the Transmission Control Protocol. The result of bufferbloat was increasing congestion and sporadic slowdowns.[38] What's most frightening about bufferbloat is that it was hiding in plain view. Gettys concluded; "the issues that create delay are not new, but their collective impact has not been widely understood . . . buffering problems have been accumulating for more than a decade."[39]

What a laundry list of accidental ways smart cities might be brittle by design or oversight! But what if someone deliberately tried to bring one to its knees? The threat of cyber-sabotage on civil infrastructure is only just beginning to capture policy makers' attention. Stuxnet, the virus that attacked Iran's nuclear weapons plant at Natanz in 2010, was just the beginning. Widely believed to the product of a joint Israeli-American operation, Stuxnet was a clever piece of malicious software, or malware, that infected computers involved with monitoring and controlling industrial machinery and infrastructure. Known by the acronym SCADA (supervisory control and data acquisition) these computer systems are industrial-grade versions of the Arduinos discussed in chapter 4. At Natanz some six thousand centrifuges were being used to enrich uranium to bomb-grade purity. Security experts believe Stuxnet, carried in on a USB thumb drive, infected and took over the SCADA systems controlling the plant's equipment. Working stealthily to knock the centrifuges off balance even as it reported to operators that all was normal, Stuxnet is believed to have put over a thousand machines out of commission, significantly slowing the refinement process, and the Iranian weapons program.[40]

The wide spread of Stuxnet was shocking. Unlike the laser-guided, bunker-busting smart bombs that would have been used in a conventional strike on the Natanz plant, Stuxnet attacked with all the precision of carpet bombing. By the time Ralph Langner, a German computer-security expert who specialized in SCADA systems, finally deduced the purpose of the unknown virus, it had been found on similar machinery not only in Iran but as far away as Pakistan, India, Indonesia, and even the United States. By August 2010, over ninety thousand Stuxnet infections were reported in 115 countries.[41]

Stuxnet was the first documented attack on SCADA systems, but it is not likely to be the last. A year later, in an interview with CNET, Langer bristled at the media's focus on attributing the attack to a specific nation. "Could this also be a threat against other installations, U.S. critical infrastructure?" he asked. "Unfortunately, the answer is yes because it can be copied easily. That's more important than the question of who did it." He warned of Stuxnet copycat attacks, and criticized governments and companies for their widespread complacence. "Most people think this was to attack a uranium enrichment plant and if I don't operate that I'm not at risk," he said. "This is completely wrong. The attack is executed on Siemens controllers and they are general-purpose products. So you will find the same products in a power plant, even in elevators."[42]

Skeptics argue that the threat of Stuxnet is overblown. Stuxnet's payload was highly targeted. It was programmed to only attack the Natanz centrifuges, and do so in a very specific way. Most importantly, it expended a highly valuable arsenal of "zero-day" attacks, undocumented vulnerabilities that can only be exploited once, after which a simple update will be issued by the software's supplier. In its report on the virus, security software firm Symantec wrote "Incredibly, Stuxnet exploits four zero-day vulnerabilities, which is unprecedented."[43]

Stuxnet's unique attributes aside, most embedded systems aren't located in bunkers, and they are increasingly vulnerable to much simpler attacks on their human operators. Little more than a year after Stuxnet was uncovered, a lone hacker known only as "pr0f" attacked the water utility of South Houston, a small town of seventeen thousand people just outside Texas's most populous city. Enraged by the US government's downplaying of a similar incident reported in Springfield, Illinois, pr0f homed in on the utility's Siemens SIMATIC software, a Web-based dashboard for remote access to the waterworks' SCADA systems. While the Springfield attack turned out to be a false alarm—federal officials eventually reported finding "no evidence of a cyber intrusion"—pr0f was already on the move, and the hacker didn't even need to write any code.[44] It turned out that the plant's operators had chosen a shockingly weak three-letter password. While pr0f's attack on South Houston could have easily been prevented, SIMATIC is widely used and full of more fundamental vulnerabilities that hackers can exploit. That summer Dillon Beresford, a security researcher at (oddly coincidentally) Houston-based network security outfit NSS Labs, had demonstrated several flaws in SIMATIC and ways to exploit them. Siemens managed to dodge the collateral damage of Stuxnet, but the holes in SIMATIC are indicative of far more serious risks it must address.

Another troubling development is the growing number of "forever day" vulnerabilities being discovered in older control systems. Unlike zero-day exploits, for which vendors and security firms can quickly deploy countermeasures and patches, forever-day exploits target holes in legacy embedded systems that manufacturers no longer support—and therefore will never be patched. The problem affects industrial-control equipment sold in the past by both Siemens and GE, as well as a host of smaller firms.[45] It has drawn increased interest from the Cyber Emergency Response Team, the government agency that coordinates American cyber-security efforts.

One obvious solution for securing smart-city infrastructure is to stop connecting it to the Internet. But "air-gapping," as this technique is known, is only a stopgap measure at best. Stuxnet, much like Agent.btz, the virus that infected the Defense Department's global computer network in 2008, were likely both walked into secure facilities on USB sticks.[46] Insecure wireless networks are everywhere, even emanating from inside our own bodies. Researchers at the security firm McAfee have successfully hijacked insulin pumps, ordering the test devices to release a lethal dose of insulin, and a group of computer scientists at the University of Washington and University of Massachusetts have disabled heart-defibrillator implants using wireless signals.[47]

These vulnerabilities are calling the entire open design of the Internet into question. No one in those early days of ARPANET ever imagined the degree to which we would embed digital networks in the support systems of our society, the carelessness with which we would do so, and the threat that malevolent forces would present. Assuring that the building blocks of smart cities are reliable will require new standards and probably new regulation. Colin Harrison, IBM's smarter-cities master engineer, argues that in the future, "if you want to connect a computer system to a piece of critical national infrastructure it's going to have to be certified in various ways."[48] We'll also have take stronger measures to harden smart cities against direct assault. South Korea has already seen attacks on its civil infrastructure by North Korean cyber-warriors. One strike is believed to have shut down air traffic control in the country for over an hour.[49]

Nothing short of a crisis will force us to confront the risk of smart cities' brittle infrastructure. The first mayor who has to deal with the breakdown of a city-scale smart system will be in new territory, but who will take the blame? The city? The military? Homeland security? The technology firms that built it? Consider the accountability challenge Stuxnet poses—we'd likely never have known about it were it not for its own bug. Carried out of Natanz by some unsuspecting Iranian engineer, the worm failed to detect that it had escaped into the open, and instead of deactivating its own reproductive mechanisms, like a real virus it proliferated across the globe.[50]

Bugged

When sensors are used without our knowledge or against our will, they become instruments of surveillance. Most of the sensors to create a seamless snooping system are already in place, but the data—credit-card transactions, passport scans at borders, e-mails, and phone calls—are held by a scattered array of organizations. Linking it all together, sifting through it and assembling dossiers is, for government intelligence agencies and law enforcement, a killer app for smart cities.

If that wasn't yet clear, it became abundantly so when Vice Admiral John Poindexter returned to public service in 2002 to launch Total Information Awareness (TIA), the Pentagon's effort to data-mine the global war on terror. Poindexter was an odd choice to head the program—his conviction in 1990 for lying to Congress about the Iran-Contra affair, while later reversed, subjected the program to increased scrutiny by civil-rights watchdogs.

Total Information Awareness was just as ominous as it sounds. At its heart was an effort to build what the Defense Department described as a "virtual, centralized, grand database" of government records, commercial transactions, and intercepted private communications. This data would be used to compute risk profiles of foreign visitors and American citizens alike, and mine it all for patterns of terrorist activity. Under intense scrutiny over another aspect of the program—a virtual market for trading predictions about geopolitical events, which people believed terrorists might use to profit from their own crimes—Congress defunded the project just as it was ramping up in 2003.[51]

In the meantime, however, much of the technology agenda of Total Information Awareness has been implemented by other governments and private firms around the world. In an odd geographic reconfiguration of power and control, every move, transaction, and message of city dwellers is now secreted away by fiber optics to become feedstock for pattern-matching algorithms grinding away in exurban server farms. Once havens of anonymity, big cities are fast becoming digital fishbowls. But while TIA's grand database sought to find traces of terror cells in big data, the real value of all this covert watching is more mundane. It's about money.

It starts in our pockets. Mobile devices, like the iPhone, keep a running record of where we've been. Apple quietly disclosed this practice in 2010, but it didn't make headlines until a year later when security experts Alasdair Allan and Pete Warden created a tool for users to easily access and map it. The data wasn't just comprehensive and detailed; it was unencrypted and copied to every machine you synced with.[52] Owners of non-Apple smartphones smirked, but a half-year later, another scandal broke over the widespread use of Carrier IQ software on other manufacturers' devices. And Carrier IQ didn't just track location. As documented by Trevor Eckhart, a systems administrator living in Connecticut, it also tracked dropped calls and every single click and keystroke made by the owner.[53] Wireless companies claimed this data was indispensable for troubleshooting technical problems, but privacy watchdogs were stunned by its level of detail.

Most phones allow you to turn off location tracking, but mobile devices can also be used to track us passively, without our knowledge or consent, through systems that monitor the unique wireless beacons phones send out as they communicate with nearby towers. One such system, called FootPath, is sold by Portsmouth, England–based Path Intelligence. As the 2011 holiday shopping season approached, American consumers were surprised to learn Forest City Commercial Management, an operator of shopping malls, had deployed FootPath to track shoppers in California and Virginia.[54] To map our movements, FootPath relies on a carefully placed array of listening posts to track mobile devices as they wander around a building. By triangulating the beacons sent by our phones to nearby cell towers, our location can be pinpointed with an accuracy of "a few meters" (the company doesn't publicly specify beyond that), enough to know how you move from store to store, your "dwell time" spent inside, the sequence of shops visited, and even movements between sections inside large department stores. FootPath probably gets paid on both sides—it can sell the demographics to retailers, as well as to mall operators who can use it to negotiate higher rents. Other than a sign at the mall entrance inviting shoppers to opt out by turning off their phones, the system is invisible, passive, and undetectable. Google and Nokia are also working on their own indoor positioning systems, and wireless chip manufacturer Broadcom is building features to support it in its products. "Acting like a glorified pedometer," one tech blogger explains, "this Broadcom chip could almost track your movements without wireless network triangulation." Using a navigational technique known as "dead reckoning" (the same way your car updates your position in a tunnel when it can't receive signals from GPS satellites), "it simply has to take note of your entry point (via GPS), and then count your steps (accelerometer), direction (gyroscope), and altitude (altimeter)."[55]

Despite Congress's objections to Total Information Awareness, law enforcement is finding the honeypot of personal data wireless carriers is accumulating irresistible. According to information filed in response to a congressional investigation in 2012, AT&T alone received over 260,000 requests for subscriber location data from American law enforcement organizations in 2011, compared to just over 125,000 in 2007—more than doubling while the company's subscriber base grew by less than 50 percent over the same period. The company now employs more than one hundred full-time workers to respond to law-enforcement requests.[56] As the *New York Times* reported, "the widened cell surveillance cut across all levels of government—from run-of-the-mill street crimes handled by local police departments to financial crimes and intelligence investigations at the state and federal levels."[57]

In many parts of the world, mass urban surveillance is overt and often welcomed. In recent years Chinese authorities have implemented two of the largest urban surveillance projects ever attempted. In November 2010, without public objection, the city of Chongqing launched an effort, inauspiciously dubbed "Peaceful Chongqing," to install some five hundred thousand video cameras that will soon watch every street corner and plaza in the giant metropolis, keeping an eye on more than 6 million people.[58] No doubt the municipal government (under the thumb of law-and-order mayor Bo Xilai, who has since been removed from power on suspicion of corruption) was inspired by the success of a similar network of over twenty-five thousand cameras in the Arab Emirate of Dubai that revealed frame-by-frame how foreign assassins infiltrated the Al Bustan Rotana Hotel to kill Hamas leader Mahmoud al-Mabhouh in January 2010. From the first known use of closed-circuit television cameras to monitor crowds in London's Trafalgar Square during a state visit by the king and queen of Thailand in 1960, urban video surveillance has come a long way.[59] The Brookings Institution calculates that today it would cost $300 million in storage capacity to capture a year's worth of footage from Chongqinq's vast camera network. But by 2020, thanks to the steady decline of cost for digital storage devices, that figure could be just $3 million per year. "For the first time ever," they warn, "it will become technologically and financially feasible for authoritarian governments to record nearly everything that is said or done within their borders—every phone conversation, electronic message, social media interaction, the movements of nearly every person and vehicle, and video from every street corner."[60] What's worse is the active involvement of American firms like Cisco, which is supplying the city with network technology optimized for video transmission for an undisclosed sum.[61]

Other Chinese cities have their own ideas about tracking citizens' phones and, as with so many things, intend to do it on a scale unmatched by any nation. In March 2011 city officials in Beijing announced that a comprehensive program for tracking the populace's 17 million mobile phones would be put in place for real-time traffic management. Perhaps reflecting the greater global scrutiny of China's new would-be world capital, or shifting values among its new middle class, the Beijing project was greeted by Chinese newspapers as an invasion of privacy.[62]

The extent to which mass urban surveillance will be tolerated in smart cities will differ around the world. Government, with varying degrees of citizen input, will need to strike a balance between the costs of intrusion and the benefits of early detection. In the European Union, for instance, strong legal protections for the privacy of personal information draw clear lines (for companies at least) on how data can be collected, stored, and reused. In much of urban Asia, historically speaking, privacy is a new luxury. The differing reactions to surveillance in China's wealthy coastal cities and its industrializing core are as different as what you'd expect between San Francisco and Boise. Governments will play their hands differently. Autocratic elites like those that rule much the Persian Gulf region look at surveillance and data mining as a force multiplier that gives them leverage over terrorists, criminal organizations, oppressed minorities, and guest workers. Americans seem resigned to muddle through, leaving the courts to settle conflicts over digital surveillance and privacy on a case-by-case basis.

Mass surveillance, designed to protect smart cities, may actually put their residents at great risk. Once assembled, stockpiles of personal data are a honeypot for criminals. Theft of personal data is now endemic and epic in scale—just a single breach of security in April 2011 led to the theft of over 75 million user records from the Sony PlayStation Network, an online community for computer gamers. The stolen data included users' names, addresses, passwords, credit-card numbers, and birth dates.

Even the surveillance specialists seem overwhelmed. At the peak of the Carrier IQ scandal, information surfaced that much of the tracking was being done by extra code inserted by phone manufacturers. Carrier IQ's executives were flummoxed to find their software had been hacked by their own customers. "We're as surprised as anybody to see all that information flowing," remarked Carrier IQ marketing director Andrew Coward.[63] As *Slate*'s Farhad Manjoo put it, "these innocent explanations are exactly why you should worry that your phone is secretly invading your privacy: Between the manufacturer, the carrier, the O.S. maker, and all the other hands that touched your phone, there are more than enough opportunities to add software that overreaches, either benignly or with some malicious purpose."[64]

Private surveillance systems that connect to the cloud are open targets too. Trendnet, a company that provides surveillance solutions for homes and businesses, was compromised in early 2012. Links to live streams from thousands of its cameras were posted to hacker sites. As one report described the breach, "Some of the more interesting camera feeds included a laundromat in Los Angeles, a bar and grill in Virginia, living rooms in Korea and Hong Kong, offices in Moscow, a Newark man watching the football game in a Giants jersey, and the inside of a turtle cage."[65]

If all of this summons thoughts of George Orwell's fictional dystopia *1984*, you're not alone. In an August 2011 ruling that blocked the US government's attempted warrantless seizure of subscriber location data from Verizon Wireless during a criminal investigation, federal judge Nicholas Garaufis wrote, "While the government's monitoring of our thoughts may be the archetypical Orwellian intrusion, the government's surveillance of our movements over a considerable time period through new technologies, such as the collection of cell-site-location records, without the protections of the Fourth Amendment, puts our country far closer to Oceania than our Constitution permits."[66]

Take Cisco's vision of Songdo (and by extension the new China), an urban civilization powered by ubiquitous two-way video screens, and fold in the latest in biometrics. It would be hard to design a more flawless replica of Orwell's "telescreen," which pumped out propaganda while watching vigilantly for hints of dissent. As Orwell wrote in *1984*, "It was terribly dangerous to let your thoughts wander when you were in any public place or within range of a telescreen. The smallest thing could give you away. A nervous tic, an unconscious look of anxiety, a habit of muttering to yourself—anything that carried with it the suggestion of abnormality, of having something to hide. In any case, to wear an improper expression on your face . . . was itself a punishable offense. There was even a word for it in Newspeak: *facecrime*, it was called."[67] Peaceful Chongqing is just a warm-up for Cisco. The market for surveillance products in China is growing at double-digit rates.[68] It's a future where police, bureaucrats, employers, and hackers may look out from every screen we look into.

We'd like to think of smart technology as a benevolent omniscience, always acting in our interests. That's certainly the pitch by technology giants, governments, and start-ups alike. But the proliferation of surveillance mechanisms isn't an accident. Governments, who ought to be the ones drawing a line to protect us, can't keep themselves away from the stuff. It's so tempting that even after Congress shut down the Pentagon's Total Information Awareness program in 2003, the National Security Agency went on to build a clandestine version of the same monitoring system, even borrowing some of TIA's own prototype technology.[69] As the Brookings report on Peaceful Chongqing concluded, "Governments with a history of using all of the tools at their disposal to track and monitor their citizens will undoubtedly make full use of this capability once it becomes available."[70] The study purported to deal only with authoritarian states, but it might just as easily have included the United States.

In our rush to build smart cities on a foundation of technologies for sensing and control of the world around us, should we be at all surprised when they are turned around to control us?

Thinking About the Unthinkable

Every day, we are doubling down on a bet that technology will solve the problems of twenty-first-century urbanization, from traffic to crime to energy. But what if smart cities do turn out to be buggy, brittle, and bugged? It's unthinkable. But it may come to pass anyway.

Considering worst-case scenarios is painful, but it can lead to drastically different conclusions and actions.

Consider US strategy during the Cold War. In the early 1960s, the nuclear arms race between the United States and the Soviet Union entered a new and alarming phase. At first, American strategy was based on deterrence. By matching Soviet buildup, the United States could ensure that nuclear war would cause such total annihilation that it would be an unthinkable option for the enemy. But some thinkers, led by Herman Kahn at RAND, didn't buy the "mutually assured destruction" doctrine. In a controversial 1962 treatise, titled *Thinking About the Unthinkable*, published after he left RAND to found his own group, the Hudson Institute, Kahn argued that not only was a nuclear war winnable, but "the living would not envy the dead" as conventional thinking held.[71] Many, if not most of the population, would survive. Life would continue. Kahn's simple point—that the overly simplistic assumption of total annihilation prevented the consideration of other possible scenarios—had huge impacts on US strategy. A good defense against nuclear weapons was suddenly as important as using them offensively. If the United States could show that it could survive a Soviet sneak attack and launch a counterstrike, deterrence would be more effective.

Thinking about the unthinkable dictated a whole new approach to building cities. By concentrating population, infrastructure, and industrial capacity in nice, big, juicy, megaton-sized targets, they had become a liability in the nuclear age. As early as 1950, none other than the father of cybernetics, Norbert Wiener, wrote in *Life* magazine, "The decentralization of our cities on the spots on which they stand, plus the release of our whole communications system from the threat of a disastrous tie-up, are reforms which are long overdue. . . . For a city is primarily a communications center, serving the same purpose as a nerve center in the body."[72] While suburbanization was driven by broader economic and technological forces, defense planners certainly welcomed and encouraged the decentralization of population.[73] The federal government was much less subtle with businesses, intensively studying and promoting "industrial dispersion" throughout the 1950s.[74]

Today, our own doomsday scenario is also man-made. To avoid irreversible climate change, the International Energy Agency estimates that we need to stabilize the concentration of carbon dioxide in the atmosphere below 450 parts per million. At current rates of greenhouse-gas emission, the point of no return will arrive sometime around 2017. After that, global warming of more than 2 degrees Celsius can still be avoided, but it will cost four to five times as much as we extensively retrofit old, inefficient power plants and infrastructure.[75]

Economist Edward Glaeser of Harvard University sees cities as green alternatives to help stabilize emissions. That makes sense in America, where higher population density would dramatically slash the energy we waste through sprawl. Residents of transit-dependent Manhattan have the lowest per capita carbon output of any American community, argues David Owen in *Green Metropolis*. But for the newly emerging global middle class, even a Manhattan lifestyle represents an enormous increase in energy consumption. We have to figure out how to support a middle-class urban existence with only the carbon footprint of a villager if we are to keep global emissions from ballooning. Even Manhattanites will have to clean up their act.

The technology giants we saw in chapters 1 and 2 are pitching smart technology as the solution to this Gordian knot. In their view, there is no alternative. Smart cities are the best last hope for our survival as a species. But there are at least five different ways that we might not make it. Each is as unthinkable as the next.

First, smart technology might not deliver enough efficiency. The improvement needed to stabilize carbon dioxide emissions are "neither trivial nor impossible," according to a 2007 United Nations Foundation report. But they are certainly not a sure thing. Global energy demand grew 50 percent from 1980 to 2005, and is expected to rise another 50 percent through 2030. To stabilize atmospheric carbon dioxide below an even less ambitious target of 550 parts per million, the G8 group of industrialized nations would have to double their average annual rate of increase in energy efficiency to 2.5 percent right now, and maintain that pace of improvement through 2030.[76] But even in cities that are aggressively pursuing efficiency, progress is slow. As we saw in chapter 5, even in Amsterdam—widely regarded as a global leader in sustainability—emissions are still *rising* by one percent annually.[77] In the worst case, more efficient smart infrastructure will actually work to hold down the price of energy and stimulate even more consumption—what economists call the "rebound effect."[78]

Second, smart technology might turn out to be less effective in curbing energy use, yet highly effective for reducing traffic congestion and fighting crime. Although cities would become more appealing places to live as quality of life improved, and in America, this might help with the energy problem indirectly, by enticing people back from the suburbs to denser communities, in the developing world it could speed up the growth of megacities powered by today's dirty energy technologies. That would be an economic success story of epic proportion, but a global ecological disaster. Imagine a smart Johannesburg suddenly free of crime and booming, absorbing millions of migrants from sub-Saharan Africa into a ramshackle infrastructure of dirty minibuses and smoky coal- and dung-fueled stoves.

A third doomsday story goes like this—we do crack the code of sustainable design and bring the needed technologies to market, just not in time. Building a smart city is not like buying a mobile phone or installing a software update; it's more like open-heart surgery. Even in Singapore, with its long and proven tradition of technocratic planning, smart infrastructure projects move at a snail's pace. Since the 1970s, city managers had used a paper-based system of tolls to control access to the congested city center.[79] But when it came time to digitize the system in the 1990s, it took fully twelve years to implement the change. London's congestion-pricing system took just a year to implement after the green light was given in February 2002. But that was after thirty-eight years of deliberation. The idea was first proposed in 1964.[80]

The fourth way things could go wrong is economic stagnation. If the malaise of the developing world is too much growth, for the rich cities of the global north it may be too little. If smart technology doesn't improve our productivity, we might not be able to pay for further improvements in energy efficiency. Many hope for a return to the "New Economy" of the late 1990s, when the United States experienced a historic period of rapid increases in productivity driven, we thought, by advances in information technology. But recent research has questioned this explanation. Robert Gordon at Northwestern University notes that the greatest productivity gains from information technology during that expansion were in manufacturing of durable goods, and that it was small in historical terms. "Computers and the internet do not measure up to the Great Inventions of the late nineteenth and early twentieth century," he argued, "and in this sense do not merit the label of 'Industrial Revolution.'"[81] Furthermore, these gains soon disappeared and most developed economies saw little productivity growth during the 2000s. Any thought of an economic boom as we upgrade cities with more of the same may be premature.

In our final unthinkable future only the wealthy thrive, retreating to smart enclaves sustained by captured resources managed solely for their own benefit, or traded at onerous rates with the poor. This scenario is already the norm across much of the developing world, where the poor have less access to clean water, healthy food, and basic sanitation, and pay vastly higher prices for them when they do. As competition for natural resources heats up over the next century, and the impacts of climate change disrupt supplies, the rich may be able to wall themselves off from the consequences of their own overconsumption. Instead of making cities more resilient to the challenges of rapid growth and climate change, smart technology could limit the ability of poor and vulnerable communities to adapt.

Every smart city will be buggy, brittle, and bugged in its own peculiar ways. It is self-delusion to expect anything else. Thinking about the unthinkable needs to be a bigger part of our discussions about the future of the city, the role technology should play, and how we manage the risks that come along with it.

A half-century ago, motorization promised to save us from the environmental crises of the day—the crowding of cities, and their lack of fresh air and green space. But imagine if we had stopped to think about the unthinkable. Could we have anticipated smog, sprawl, dependence on foreign oil, childhood obesity, and global warming? We will never know if these negative impacts could have been avoided, but it would not have cost much to try. We might have even avoided the very unintended consequences we now invent smart technologies to undo.

10

A New Civics for a Smart Century

We have seen that putting the needs of citizens first isn't only a more just way to build cities. It is also a way to craft better technology, and do so faster and more frugally. And giving people a role in the process will ultimately lead to greater success in tackling thorny urban problems and greater acceptance of the solutions smart cities will offer. Oscar Wilde once wrote, "At present machinery competes against man. Under proper conditions machinery will serve man."[1] It is up to us to create the right conditions. But if we want to put people first, where do we begin?

I believe we need a new set of principles to guide us. These principles need to build not only on our growing scientific understanding of cities and how technology shapes and is shaped by them, but also a broader appreciation of the human condition and how it is changing in this first predominantly urban century. To put it simply, we need science, but we also need culture to chart the way forward.

In chapter 3 we saw how the roots of modern city planning grew from Patrick Geddes's evolutionary understanding of cities and his belief that the practical application of sociology was crucial to solving the fast-multiplying problems of industrial-era cities. Geddes would no doubt approve of how today's smart-city builders are applying technology to urban challenges and seeking to develop a new, rigorous empirical science of cities. But he also understood the limits of science, and the need to view cities with eyes that see not only facts, but wonder as well. As biographer Helen Meller wrote, Geddes believed that "the city had to be seen as a whole, not as an amalgam of disparate elements each requiring specific treatment. . . . Seeing the city as a whole however, was not straightforward; it required a special combination of science and art. Scientific facts, observations made in a systematic manner, combined with an artistic understanding based on cultural criteria, together made a new subject Geddes called 'civics.' It was only possible to study this subject in a specific context and therefore the beginning of such a study had to be a practical social survey."[2]

Geddes recognized that a thorough knowledge of culture—the creative social expression of humanity in a particular local setting—was necessary to understand what science could not explain. Today, as computers do more and more of the work of observing cities for us, we must redouble our efforts to see those intangible aspects of urban life they may never be capable of measuring. Without this more holistic lens on the city, it will be impossible to recognize problems, design appropriate solutions, and engage citizens to participate in their implementation.

Yet evidence that we are moving in the wrong direction is everywhere. As we saw in chapter 2, visionary computer scientist David Gelernter was deeply conflicted about the death of Romantic thought under the relentless scrutiny of mirror worlds—technological contraptions not unlike the ones that IBM has engineered in Rio de Janeiro. When I think about how Mayor Eduardo Paes's remote-control city reduces the people of the favelas to a stream of data, the words of E. E. Cummings, who railed against the mechanization of a life ruthlessly measured, come to mind:

—bring forth your flowers and machinery: sculpture and prose
flowers guess and miss
machinery is the more accurate, yes
it delivers the goods, Heaven knows

Smart cities designed by corporations will deliver, indeed. But what? A landscape of automated cookie-cutter urbanism that doubles down on industrial capitalism and inevitably crushes our souls? Again, a few lines down, towards the end of the poem, Cummings draws our attention to the stakes:

who cares if some oneeyed son of a bitch
invents an instrument to measure Spring with?[3]

What we stand to lose from this urge to wire up the planet with sensors is, ironically, itself immeasurable. I wonder if it's time to jet down to Rio, pull the plug on the Intelligent Operations Center, and put the boys of Projecto Morrinho, with their Lilliputian model of the city, in charge instead.

Failure to put people at the center of our schemes for smart cities risks repeating the failed designs of the twentieth century. Only this time, the stakes are much higher, because by the end of this century, with as much as 80 percent of the world's population already living in urban areas, there will be few cities left to build. As economist Paul Romer points out, "in the lifetimes of our children, the urbanization project will be completed. We will have built the system of cities that their descendants will live with forever."[4] Walk amid new Songdo's shiny new towers, and one thing is abundantly clear—it is a twenty-first-century update of the Garden City. Jane Jacobs was right about the pointlessness of model cities designed by professional planners. But this is where we are placing our bets.

Until now, smart-city visions have been about controlling us. What we need is a new social code to bring meaning to and exert control over the technological code of urban operating systems. We need a new civics for the smart city that takes what we know about making good places as well as good technology, and shows us how to put them into practice. Only a sound set of guidelines will allow the designs for smart cities to emerge organically and to be shaped by the desires and choices of the people who must live in them.

In these closing pages, I offer a set of tenets that we can use to build this new civics. They are my distillation of the crucial design, planning, and governance principles we must uphold to build smart cities that are human-centered, inclusive, and resilient. It is unavoidably incomplete—the fast-changing nature of both cities and computing makes it impossible to capture all of the important issues. We might heed the words of the late William Mitchell, the former dean of MIT's School of Architecture and a pioneering thinker on smart cities, who wrote, "our job is to design the future we want not to predict its predetermined path."[5] This is, I hope, the beginning of a new phase in our collective conversation about how to do that.

Opt In to Smart

The commercial success and cultural ascendance of the Internet lends an air of inevitability to the idea of smart cities. But are we too eager to ask engineers to solve every urban problem? The technology industry's hard sell on smart depends on this. But only the company towns of the twenty-first century will see technology as the end goal. The first tenet of our new civics is that we should never default to smart technology as the solution. It's tempting to think that new gadgets always offer better solutions to old problems. But they are just another set of tools in an already well-equipped box.

One need only open up Christopher Alexander's monumental book *A Pattern Language* to understand just how big that toolbox is. The result of a decade's worth of painstaking research, it is a fascinating distillation of humanity's built legacy, describing over two hundred traditional architectural and urban design tropes from cities around the world. What *A Pattern Language* argues is that most urban design problems were solved long ago by ancient builders. We have but to borrow from our ancestors, and many problems can be adequately addressed simply by conventional design.

Instead, however, we are creating technological bandages to fix flaws in the poor designs of mass-produced cities. Consider the distribution of commerce and industry. Alexander's Pattern 9, "Scattered Work," described the network of small workshops intermixed with homes that's typical in cities that have grown organically. Scattering work integrates the social and economic life of cities, provides opportunities for young people to learn about work, enhances walkability, and reduces the commuting burden on transportation systems. Yet in the world's rapidly urbanizing countries these traditional forms, and their fine-grained mix of uses and building types, are being bulldozed to make way for single-use districts. In a headlong rush to modernize, Chinese cities are repeating one of the West's worst mistakes of the twentieth century, and doing so on an epic scale. But, as Cisco pitched at the 2010 World Expo, technology can undo the damage—ubiquitous videoconferencing will patch Shanghai's fractured landscape back together. However, this strategy can only postpone the inevitable structural changes needed to make these modern designs stand the test of time as well as Alexander's patterns have.

We needn't all become Luddites overnight. Treat smart as an add-on, an upgrade, and not the end itself. The best thing about smart technologies is that you don't have to clear-cut your existing city to make way for them. But ask the hard questions: What new solutions do smart technologies really enable? Where do they enhance existing solutions? Most important, where do they interfere and create new problems of their own? You can also future-proof conventional designs for smart retrofits later on. When you replace street lamps, provide a mounting point for whatever wireless or sensor technology comes next. When you dig up streets, lay conduit for future broadband lines. Whatever they may be made of, there will be powerful economic reasons to squeeze them into the same slots, just as fiber optics followed the paths laid down by the telephone and telegraph wires that preceded them. When you create urban software, make it simple, modular, and open source. Anytime you generate a new data stream, document and archive it as openly as you can.

Plan for life cycles—it's just as important to clear out old technology when you bring in the new. Cities that cling doggedly to a single technology are destined to become obsolete when the next shift occurs. What has made Alexander's patterns so persistent is their ability to evolve as foundations for new technologies and human activities.

Roll Your Own Network

A century ago, cities all over the world realized that universal access to electric power meant taking over the business themselves. Power companies had cherry-picked the best customers and most profitable districts, depriving marginal and outlying areas of the benefits of network access. Today, many cities are realizing that similar economics apply to broadband. Throughout Europe, cities such as Stockholm, Amsterdam, Cologne, and Milan have invested in public broadband infrastructure, dramatically increasing speeds and lowering costs to residents and businesses.

But, as we saw in chapter 7, state governments in America prohibit communities from building their own public broadband networks. Back in 2005, when Philadelphia was fighting for its wireless future in the Pennsylvania state legislature, US Federal Trade Commission member Jon Leibowitz told a gathering of city officials that "local governments have long been laboratories of experimentation. If they want to give their residents affordable Internet access, they should be allowed to try without being foreclosed by federal or state laws—or by cable and telephone interests."[6] Because of the restrictions enacted during that era, only about 150 communities in the United States have built public fiber-optic networks, far shy of the some 3,300 municipalities that are in the electric power business.[7]

But early movers like Chattanooga, Tennessee, which authorized its municipal power company to expand into telecommunications in

2008, show how productive an investment fiber is. The city is saving on telecommunications charges, the power authority has dramatically reduced outages through deployment of smart power-grid technologies that connect through the fiber network, and businesses have seen dramatic price drops in ultra-high-speed Internet service. Claris Networks, a local cloud services firm based in nearby Knoxville, moved jobs to Chattanooga, where its connectivity costs dropped by 90 percent.[8]

The telecommunications industry's arguments against public broadband utilities ring hollow. They vilify broadband as a financial quagmire for cities, yet by 2009 municipal fiber networks on average captured over half the market within just a few years, well above the 30 to 40 percent needed to break even. Some were projected to pay off their construction bonds early, and not a single one had failed.[9] Even the pro-market Organisation for Economic Co-operation and Development (OECD), the club of developed nations, endorses the approach, arguing that "Municipal networks can play an important role in enhancing competition in fibre networks."[10]

Community-owned broadband is one of the best investments a smart city can make. It creates a vital infrastructure for information-intensive industries, and it opens the door to new opportunities for human and social development through remote learning and immersive multimedia communications. More importantly, it puts the city in control of its own nervous system, giving it tremendous bargaining power over any private company that wants to sell smart services to the city government or its businesses and residents. By putting control over many aspects of management under local jurisdiction, community-owned networks also render moot the struggle over two important telecommunications policy issues—net neutrality, which seeks to prevent ISPs from restricting user access to content and applications, and making Internet access a human right, in accordance with a 2012 United Nations declaration. Cities could simply decree their broadband networks open and free, to both content providers and citizens without the financial wherewithal to pay for a broadband connection.

Public-private partnerships like the one that Philadelphia struck with EarthLink are too beholden to short-term market forces to work over the long haul. But many creative mechanisms for funding these networks are in the works. Municipal bonds, like a residential mortgage, allow the time horizon for return on investment to be stretched to match the useful working lifetime of the infrastructure. Through the Gig.U partnership, universities across the United States are stepping up to extend campus networks into surrounding communities. The town of Sandy, Oregon, requires real estate developers to extend the city's public fiber grid into new developments on virgin land, under the same subdivision regulations that now require them to build roads, sewers, and water mains.[11] Some communities are beginning to experiment with crowdfunding local broadband projects.

In the poorest parts of the world, more than just local fiber networks are needed—the entire cloud infrastructure that rich nations enjoy needs to be created from scratch—as we saw in Moldova (chapter 6) where a World Bank grant helped create a "g-cloud" that powers the national government's online services and internal information systems. By underwriting a large portion of the cost of a nationwide cloud infrastructure, the g-cloud will reduce the cost and expand the quality of computing services for local businesses. In lieu of grants, poor nations can justify such investments by creating shared infrastructure underwritten by military, law enforcement, and emergency-response users.

Build a Web, Not an Operating System

In the race to prescribe how the various pieces of smart cities will talk to each, there is a growing buzz about the need for an "urban operating system." Living PlanIT, the London-based software company that's building a research park for smart-city technology in the hills outside the city of Porto in Portugal, even claims a trademark on the term.

For personal computers and mobile devices, the operating system is an essential suite of software that does the heavy lifting of routine, common functions like opening and closing windows on a screen, reading keyboard input, writing to the disk—so that every new program doesn't have to reinvent the wheel. An urban operating system would handle tasks like processing your payment for a taxi fare, trafficking road sensor readings up to a server in the cloud, or verifying a residents' identity when they approach the door of their home. As bits of the smart city interact with each other, the urban operating system will broker the exchange.

For engineers, the benefits of urban operating systems are clear—faster and cheaper application development. But for business strategists, a single operating system for the city has only one purpose—to make the entity that designs it indispensable. Whoever owns this layer of proprietary protocols and infrastructure will truly hold the keys to the city. As one of Living PlanIT's executives has said publicly, the "urban operating system will control everything that happens in the city."[12] But the precedent of companies exploiting dominance in personal-computer operating systems should sound alarm bells in City Hall. Already, Living PlanIT is more focused on creating cozy relationships with technology companies whose products will plug into its operating system. Its relationship with Cisco and McLaren, a sensor manufacturer, looks like the notorious Microsoft-Intel alliance, the "Wintel" duopoly that dominated desktop computing for decades. And for years, Microsoft exploited undocumented features in the Windows code base to make its highly profitable Office software work better than competing rivals. Smart-city monopolists will design similar backdoors for their own profit.

The obvious alternative to an urban operating system is the Web and an organically evolved set of open standards and software that anyone can build on. Andrew Comer, a partner at engineering giant Buro Happold, argues, "In an ideal world, we would have common, open-source platforms that can accommodate all of these systems, and manage the transfer of information between them all. It would be more democratic, create more opportunity for competition, and make it easier for new players to bring new products to market."[13]

It is in the long-term interest of industry to map the success of the Internet and open-source software onto the city. Some of the big players are starting to get this, most notably IBM, which long ago embraced open-source software. Putting such a framework in place means mapping out the essential minimal components required to share data, process transactions, and secure critical systems. It would be a huge step toward realizing a smart city that, in Christopher Alexander's view, would look more like a lattice than a tree.

Establishing the right standards will take time, but as we saw in chapter 3, this approach has proven highly effective at driving innovation in Internet technology. And for now, the lack of standards is slowing the adoption of smart technology by making it harder for cities to combine their efforts. As Code for America founder Jennifer Pahlka asks, "what are the standards that will allow us to collaborate without discussing it?"[14]

A truly citizen-focused urban operating system should recognize, as MIT's Carlo Ratti says, that "people are the ultimate actuators of cities."[15] With greater openness and flexibility, a Web-like operating system for cities would give developers and even users the ability to design new solutions. A web of smart urban things and services will reinforce the sociability that makes cities thrive. Instead of being centralized, many vital services could be left to the social networks of small communities. A corporate operating system, by contrast, may save on the lighting bill and keep the crooks out, but in the process it could sap the vitality of the community it was trying to protect in the first place.

Extend Public Ownership

Even if one firm doesn't capture an entire city's smart infrastructure by controlling its operating system, critical pieces will inevitably be privatized. The global recession has decimated municipal ledgers everywhere. Under the benevolent guise of public-private partnerships, financiers offer capital and technology in exchange for exclusive rights to operate urban infrastructure. The most shocking instance of this occurred in 2008 when Chicago tendered a seventy-six-year lease of its thirty-six thousand parking meters to a firm backed by the government of Abu Dhabi for a $1 billion balloon payment. With cities struggling to invest in even basic infrastructure, there is little appetite for costly smart systems. But industry is getting creative. In 2012, for instance, IBM partnered with Citibank to set up a $25 million loan fund to finance smart parking systems for American cities.[16]

But what companies really lust after is our big data.

The first sign of the struggles to come showed up in San Francisco. In the early 2000s, the city's Muni transit system contracted with NextBus, a firm that provided vehicle-tracking technology, to create an arrival-time information service on its website and in transit stations. But in 2009, when civic hacker Steven Peterson launched Routesy, an iPhone app that pulled arrival times from the transit agency's website, Muni made an unpleasant discovery. It didn't own the results of the arrival predictions generated by NextBus's algorithms. In 2005, in a near-death financial crisis, NextBus had sold those rights in a fire sale to a shell company set up by one of its founders. San Francisco could post the arrival predictions on its own website, but anyone who wanted to use them for other purposes had to pay. Luckily, the issue was resolved in the city's favor when the company's contract came up for renewal later that year.[17] But as the open-data movement grows, cities everywhere are taking another look at their agreements with technology vendors and service providers.

A handful of cities, as we saw in Zaragoza, are eager to take on an expanded role as stewards of citizens' sensitive private data. They see decisions about how, where, when, why, and on what terms to share, make public, or otherwise reuse this data as important matters of public policy. They are the exception. Most local governments, especially risk-averse and fiscally constrained ones in the United States, will shun this enormous responsibility. They lack the capacity to even negotiate controls over the data streams generated by their citizens as they interact with private vendors' technologies. Watchdog groups will need to step in and identify where the crucial conflicts lie. (And in fact, the Electronic Frontier Foundation is doing just this on behalf of a number of transit agencies being sued by another transit-arrival patent troll, Luxembourg-based ArrivalStar).[18] Cities will need regular audits, perhaps conducted by a chief privacy officer or chief data officer charged with extending public control over government- and citizen-generated data.

An intriguing option is to hand off this data to a trust equipped to manage it on behalf of citizens, covering its costs—and possibly generating a revenue stream for the city—by licensing the data. A growing number of start-ups and open-source projects, like the Personal Locker project started by Jeremie Miller, are exploring ways for individuals to control and even pool their private data to trade with companies. (As the creator of Jabber, the dominant global protocol for instant messaging, Miller has a proven knack for standards.) Others are developing the technologies to aggregate and store hyperlocal data. In Brooklyn's Red Hook neighborhood, the New America Foundation's Open Technology Institute has deployed a community mapping system called Tidepools that runs off local servers instead of the cloud. Institutionalizing this infrastructure at a community scale would give cities the ability to dictate when and how citizens' data is used.

Regardless of how cities choose to manage their data, they need to think more broadly and long-term about its value. Extended public ownership of the data exhaust of cities could potentially drive new business models to pay for investments in smart systems. Even today, only a handful of cities share data through a central repository. This means there is still an opportunity to design more sophisticated models for aggregating and distributing data locally generated by government and citizens alike. Chicago's CTO John Tolva sees city data as a raw material for business. "There is an economic development argument around open data," he explained to me. "It's a platform that businesses can be built upon, just as the weather industry sits on top of the National Weather Service. We could foster the growth of companies that analyze the vital signs of cities."[19] But if companies profit from data generated by cities and their inhabitants, shouldn't the community reap a share?

Extending public control over the hardware and software of smart cities will be trickier. Much of it will be privately owned and operated by outsourcing firms under contract to city governments. Cities will have financed this smart infrastructure through fees but won't own it. More troublesome, however, is that information systems that used to be packaged as products are being restructured as services delivered across the Internet—computing power is now rented rather than sold. But this business model, pushed hard by IBM, among others, is unsettlingly similar to the one Herman Hollerith imposed on the Census Office in the 1890s. For decades, IBM thrived on its usurious relationship with customers, until a 1956 antitrust action by the US government forced it to sell, as well as lease, computers and tabulators. This unbundling was critical to breaking the firm's monopoly in the fast-growing industry.[20]

The rise of cloud-computing also raises other tricky questions for smart city governments. The first is about jurisdiction. As the servers that used to be housed in the basement at City Hall migrate into the cloud, cities' critical data and infrastructure will often physically reside in locations that may be outside their legal reach. For now, it's great to reap the lower costs of an infrastructure you share with other cities. But what if there is a dispute? How will you ever switch vendors when your data is sitting on a server in another country running proprietary software? The lack of standards for cloud services is equally disturbing because it makes vendors indispensable. You can't simply move to another company's technology because you'd have to rebuild all of the underlying systems while somehow trying to recover and migrate your old data. Imagine if we ran our physical infrastructure the way IBM would run our smart-city cloud. As Dom Ricci, a financial risk manager for a large international bank who tracks smart-city developments, points out, "you don't tear off the subway rails and replace them with a different gauge every time you change operators."[21]

Put simply, smart cities need to be savvy about what data and service infrastructure they own and what they give up to private interests in the cloud. As the financial pressures of even the most basic smart systems mount, the appeal of outsourcing and privatizing will grow. (Citing costs, Detroit instead simply pulled the plug on its 311 telephone hotline in 2012).[22] But the short-term savings may evaporate quickly once they are locked out of their own data and locked in to proprietary services.

Model Transparently

The most powerful information in the smart city is the code that controls it. Exposing the algorithms of smart-city software will be the most challenging task of all. They already govern many aspects of our lives, but we are hardly even aware of their existence.

As I explained in chapter 2, computer modeling of cities began in the 1960s. Michael Batty, the professor who runs one of the world's leading centers for research in urban simulation at University College London, describes the era as "a milieu dominated by the sense that the early and mid-twentieth century successes in science could extend to the entire realm of human affairs."[23] Yet after those early failures and a long hibernation, Batty believes a renaissance in computer simulation of cities is upon us. The historical drought of data that starved so many models of the past has given way to a flood. Computing capacity is abundant and cheap. And like all kinds of software, the development of urban simulations is accelerating. "You can build models faster and quicker," he says. "If they're no good, you can throw them away much more rapidly than you ever could in the past."[24]

The "most important attribute any model should have is transparency," argued Douglass Lee, the planning scholar who marked the end of that first wave of modeling in a seminal 1973 article. Ironically, while open-source software—which thrives on transparency—is playing a major role in this renaissance in urban modeling research, most models outside the scholarly community today receive little scrutiny. The "many eyes" philosophy that ferrets out bugs in open source is nowhere to be found.

The tools that have governed the growth of cities—the instructions embodied in master plans, maps, and regulation—have long been considered a matter of public record. Models ought to be dissected and put on display in the same way, to invite scrutiny from many perspectives. But it would also serve to educate the public about their own city and the tools and methods used to understand and improve it. Imagine Patrick Geddes's regional survey approach applied to a smart city. What a small leap it would be to turn Rio's Intelligent Operations Center from mayor's bunker into a living exhibition of the city, an Outlook Tower for the twenty-first century. Already, an onsite press room allows reporters to broadcast live views of the system in action. But more transparency should follow.

We shouldn't expect the most important code of the smart city to see the light of day anytime soon. Industry will closely guard its intellectual property. Government agencies will as well, citing security and privacy concerns to mask anxieties about accountability and competence (much as they do with data today).

Citizens will need legal tools to seize the models directly. The Freedom of Information Act and other local sunshine statutes may offer tools for obtaining code or documentation. The impacts could be profound. Imagine how differently the inequitable closings of fire stations in 1960s New York might have played out if the deeply flawed assumptions of RAND's models had been scrutinized by watchdogs. At the time, there was one case in Boston where citizen opposition "eventually corrected the modeler's assumptions" according to Lee.[25] Today assumptions are being encoded into algorithms into an increasing array of decision-support tools that inform planners and public officials as they execute their duties. But the prospects for greater scrutiny may actually be shrinking instead. New York's landmark 2012 open data law, the most comprehensive in the nation, explicitly exempts the city's computer code from disclosure.

Greater transparency could also increase confidence in computer models with the group most prepared to put them to work solving problems—urban planners themselves. But the modeling renaissance that Batty sees isn't driven by planners or even social scientists, but by physicists and computer scientists looking for extremely complex problems. As Batty told an audience at MIT in 2011, "Planners don't use the models because they don't believe they work."[26] In their eyes, the results of most models are too coarse to be useful. The models ignore political reality and the messy way groups make decisions. And while new software and abundant data are lowering the cost of creating and feeding city simulations, they are still fantastically expensive undertakings, just as Douglass Lee noted forty years ago.

Without addressing the trust issue through transparency, cybernetics may never again get its foot in the front door of city hall. As journalist David Weinberger has written, "sophisticated models derived computationally from big data—and consequently tuned by feeding results back in—might produce reliable results from processes too complex for the human brain. We would have knowledge but no understanding."[27] Such models will be scientific curios, but irrelevant to the professionals who plan our cities and the public officials that govern them. Worse, if they are kept under lock and key, they may be held in contempt by citizens who can never hope to understand the software that secretly controls their lives.

The benefits of transparency go beyond just unveiling the gear works of the smart city, challenging invalid or unjust assumptions and debugging code. The process of examination itself can be a constructive part of the city planning process, as we saw with IBM's foray into system modeling in Portland. "A transparent model is still about as likely to be wrong, but at least concerned persons can investigate the points at which they disagree," wrote Lee. "By achieving a consensus on assumptions, opposing parties may find they actually agree on [the model's] conclusions."[28] And the process of modeling, if done openly and collaboratively, can create new alliances for progressive change. As IBM's Justin Cook, who led the development of the system model for Portland in 2011, explains, "you start to see that there's natural constituencies that have not identified each other . . . that the people that care a lot about obesity and the people that care a lot about carbon have something in common."[29]

Fail Gracefully

In *Mirror Worlds*, computer scientist David Gelernter compared the modern corporation to a fly-by-wire fighter aircraft: "It's so fantastically advanced that you can't fly it. It is aerodynamically unstable. It needs to have its 'flight surfaces' adjusted by computer every few thousandths of a second or it will bop off on its own, out of control. Modern organizations are in many cases close to the same level of attainment—except that, when they're out of control, they don't crash in flames; they shamble on blindly forever."[30] Engineers would rather describe this state of affairs as "graceful failure." Instead of completely collapsing, the company (or a smart city) simply lumbers on at a lower level of performance. Compared to a crash, this is actually a pretty good outcome, assuming it eventually stages a full recovery.

We know that smart cities will have bugs. Even when a botched software update brings down an entire subway system, the problem can be fixed, usually quickly. But what happens during a crisis? How will the delicately engineered balance of material and information flows in smart cities, optimized for normal peacetime operation, perform under the severe, sustained stress of a disaster or war? As we saw in chapter 9, these systems routinely break down catastrophically during such events. How can we harden smart cities and ensure that when parts of them fail, they do so in controllable ways, and that vital public services can continue to operate even if they are cut off?

Big technology companies are staring to understand the need for building resilience into smart-city infrastructure. According to IBM's Colin Harrison, "because of the complexity of these systems, if you start to overload them, they may fail. But if they fail, you'd like them to fail in a soft way, so that the operation continues, the lights don't go out, and the water doesn't stop flowing. It might not be as pressurized as you'd like it to be, but at least there will still be water." It's an extension of what systems engineers call "dependable computing," a thirty-year-old set of techniques that will increasingly be applied to urban infrastructure. At the very least, like the robots in Isaac Asimov's science fiction stories whose code of conduct prevents them from hurting humans, "it protects itself against doing harm to the infrastructure it's trying to control," imagines Harrison.[31]

Cities must set high expectations for reliability as they work with industry, and create capacity for more resilient fallbacks. Meanwhile, they must prepare for the worst. This means having a clear division of authority, plans for backup controls and services, checklists for relief efforts, methods for preventing cascading failures between interconnected urban systems, and organizational capacity to cope with surprises. Many cities already conduct environmental impact assessments, intense audits of the risks of new infrastructure and development projects. Applying this kind of scrutiny to smart-technology projects would help address public concerns about reliability, as well as provide a stamp of approval for technology products, much the way the testing and certification by independent groups like Underwriters Laboratories helped instill trust in the safeness of industrial and consumer goods.

There is a dark side to graceful failure—the same precautions taken to manage an orderly shutdown of urban infrastructure might be used to do it deliberately. There is a very real potential for politics or social upheaval to trigger graduated withdrawals of public services. Many governments already have the equivalent of an Internet "kill switch" in place, as evidenced by the Egyptian authorities' shutdown of Cairo's Internet and cellular grid by coercing telecommunications providers and ISPs to disconnect during the peak of the January 2011 Arab Spring revolt. As urban dashboards like Rio's Intelligent Operations Center evolve into remote controls as well, they'll provide a new level of precision for targeted blackouts of infrastructure and services. Entire districts of the city, or even individual buildings or dwelling units, could be selectively disconnected from the grid. Even more insidious kinds of brownouts are possible too—the flow of water, power and communications to a neighborhood might be throttled back to deliver political punishment, but controlled by carefully calibrated algorithms to level off the embargo just before it provokes an organized response from the people living there.

Build Locally, Trade Globally

Where we build the technology we use in our smart cities may matter almost as much as what we build. There are few killer apps for smart cities today. But now is not the time to close off our thinking. In the coming decade each city must strive to be as good a civic laboratory as it can be, spin out its own situated software, and with luck evolve a few smart-city genes that can spread and thrive globally.

Doing this properly will mean sustaining a modest level of investment in smart-city public works over the next ten years. One possible model is the set-aside. Many cities already mandate that a small fraction (as little as one percent) of the construction budget for public buildings be spent on public art. What if we required a similar approach to smart technology? Jay Nath, San Francisco's director of innovation, proposed just such an idea on his blog in early 2012. "A new playground could experiment with intelligent lighting that operates based on time and motion," he imagined.[32] Such a regulation would need to be carefully crafted to generate innovations with high civic value. But it would create a steady market for local smart-city tech start-ups that doesn't exist today.

Every civic laboratory needs a physical and social support system for hackers and entrepreneurs to experiment within. Contests, contracts for specific apps, and networking events are critical. Open data and read/write government information systems like Open311 create opportunities for both conceptual and commercial experiments. Physical hack spaces like New York University's Interactive Telecommunications Program, Zaragoza's Center for Art and Technology, and Code for America's accelerator literally create laboratories for inventors to work on future smart-city technologies. Big private-sector infrastructure projects, like Google's Kansas City fiber grid, can mobilize resources across the board. Before Google's geeks pulled a single strand of glass, dozens of self-organizing civic initiatives sprang to life to anticipate and maximize its impact.

Building local innovation capacity isn't enough. Smart cities will need to tap into the rich international trade in urban technology. Groups like Code for America and Living Labs Global provide access to a fast-growing pool of resources, so that cities don't have to invent from scratch a tool for every project. But more of these computational leadership networks will need to be created and sustained. They must continue to evolve beyond sharing case studies and anecdotes, to cross-fertilizing actual data, models, software, hardware designs, and business models. They must provide cities with incentives to share, and designers with advice on how to build systems that can solve local problems and be reused elsewhere.

The economic potential for cities is obvious: the best way to share is to incubate businesses that can export their innovations. But it's not just other cities that will buy them. Civic labs are already having interesting spillovers into other sectors, because they are ideal settings to explore new ways of communicating and computing. Megaphone Labs, another spin-out from ITP, was originally created by Dan Albritton and Jury

Hahn as a way to play games on the massive digital screens of Times Square using touch-tone phone codes. But after struggling to find a market for the technology, the company "pivoted"—in start-up speak. Recruiting media industry veteran Mark Yackanich as CEO, Megaphone employed the same technology to turn your phone into a remote control, and launched an assault on the cable industry's stranglehold on interactive TV. This kind of experimentation in civic labs will have ripple effects on the media, culture, and industry that can create sizable economic returns.

The key will be to balance what you build, what you import as-is, and what you tailor from a borrowed template. The risk of too many bespoke inventions is a quirky local fork that reduces your ability to borrow from others. The risk of too much borrowing or standardizing around a single tool is generic design. As Phil Bernstein of AutoDesk, a maker of architectural software, has said, "I used to be able to drive around American cities and tell you what version of AutoCAD was used to design each building."[33]

The greatest risk of this approach is that cities that lack the capacity to design their own smart solutions will fall behind. Today, only a handful of cities have the capacity to develop their own technologies locally; a somewhat larger group is able to import solutions and replicate what others have done. But just as we have struggled to expand broadband networks in smaller and poorer communities, directed effort to expand access and literacy in smart city technology will be needed.

Cross-Train Designers

Inspired by Patrick Geddes's view of the region as an integrated human and natural system, New Urbanism pioneer Andres Duany developed the notion of the "urban transect" in the 1990s. A cross-sectional diagram, the transect describes the zones of ever-greater density that characterize the journey from a city center through suburbs and into the hinterlands. The transect was a tool to help designers think about the interfaces and the transitions between different parts of the built and natural world.[34] The challenge for designers of smart cities will be navigating another transect, the one that connects the physical and the virtual world. To do so effectively, they'll need to cross-train.

This cross-training will take two forms. First, they will need to heed Geddes's admonition to see cities as both scientists and artists. As Red Burns, the cofounder of NYU's Interactive Telecommunications Program, once described the curriculum's goal: "we are training a new kind of professional—one who is comfortable with both analytical and creative modes of thinking."[35] Similarly, it won't be enough to just put together teams with both planners and programmers. Smart-city designers will also need to be transdisciplinary—able to think across disciplines inside their own minds. As author Howard Rheingold describes it, transdisciplinarity "means educating researchers who can speak languages of multiple disciplines—biologists who have an understanding of mathematics, mathematicians who understand biology."[36] Architects and engineers of smart cities will need to draw on both informatics and urbanism simultaneously. There are about a dozen people in the world today who can do this proficiently. One of them, Adam Greenfield, argues that future designers of smart cities, "will have to be at least as familiar with the work of Jane Jacobs . . . as they are with that of Vint Cerf," the computer scientist widely considered to be one of the founding fathers of the Internet.[37] To be effective in getting their designs built, they will need to deeply understand smart systems and their risks and benefits, and be able to explain it all to nonexpert stakeholders.

To date, the few transdisciplinarians working on smart cities are mostly technologists or scientists dabbling in urbanism. But as a discipline, urban planning is probably better prepared to systematically cross-train its own students from the other direction. That's because planning is already connected to a hodgepodge of disciplines that offer insights on the city: engineering, economics, sociology, geography, political science, law, and public finance. Expanding its small existing connection to informatics would be easy.

The need for a broader perspective on smart systems is so clear that even those outside the field see it. Writing about the future of the ICT4D movement in *Boston Review*, Evgeny Morozov argued:

> In short, we need to be realistic, holistic, and attentive to context. Why haven't we been so far? Part of the problem seems to lie in the public's penchant for fetishizing the engineer as the ultimate savior, as if superb knowledge of technology could ever make up for ignorance of local norms, customs, and regulations. . . . Non-technologists may be more successful in identifying the shortcomings of technologies in given contexts. They may be better equipped to foresee how proposed technological solutions complement or compete with other available non-technological solutions as well as to anticipate the political and institutional backlash that can result from choices of technology.[38]

These are precisely the problem-solving approaches that urban planners use every day.

Yet even as a company like IBM boasted about a track record of two thousand smart-city engagements in 2011, it hired just a single urban planner—as far as I can ascertain, the company's first.

Think Long-Term in Real Time

At a conference in Singapore in early 2012, New York's Michael Bloomberg lamented that "Social media is going to make it even more difficult to make long-term investments" in cities.[39] As mayor, Bloomberg had pushed city agencies hard to engage the public, creating over two hundred social-media channels. But when citizens used social networks to talk among themselves, the conversation snowballed into daily referenda on his administration.

Figuring out how to harness real-time data and media to think about long-term challenges is one of the most important opportunities we must exploit. But throughout history, planners have struggled to create durable visions. Cities don't stand still, and often change in unpredictable ways. Italo Calvino captured the challenge in his novel *Invisible Cities*:

> In the center of Fedora, that gray stone metropolis, stands a metal building with a crystal globe in every room. Looking into each globe, you see a blue city, the model of a different Fedora. These are the forms the city could have taken if, for one reason or another, it had not become what we see today. In every age someone, looking at Fedora as it was, imagined a very different way of making it the ideal city, but while he constructed his miniature model, Fedora was already no longer the same as before, and what had been until yesterday a possible future became only a toy in a glass globe.[40]

In smart cities static visions will be even less durable, as both reality and our models of it change second-by-second.

If city planning is to keep up, it needs to become more of an agile, fluid process than the semi-decennial slog it is today in most cities. Michael Joroff, who studies planning and development at MIT, argues that "planning is going to be more iterative than in the past. Master plans will give way to master strategies."[41] In his view, these new visions will combine fixed, predictable elements along with placeholders that will be fleshed out later. This approach allows plans to be updated frequently to reflect changes in society, economy, and environment. More importantly, it creates an opportunity for the torrents of data produced by smart systems to inform those tweaks. Smart-city boosters herald big data's value in prediction, but in the near-term it will be far more valuable in merely decoding the detail of how past decisions actually changed the city. Planners will still make judgment calls, but they will be better informed about potential outcomes. For instance, when New York City closed Times Square to vehicles during a pedestrianization campaign, it used GPS data from taxis to both predict and verify changes to traffic patterns in the surrounding area.[42] IBM's Guru Banavar, who led the company's work on the Rio Intelligent Operations Center, sees a "feedback loop between the planning and operations of a city . . . the day-to-day activities and day-to-day successes and failures . . . can provide information historically about how the next round of planning ought to be done."[43] As Joroff explains, "Big data will inform strategy on a macro scale. We will better know about conditions and consequences of policies and actions. Ignorance will no longer be a condition or an excuse. If the political will is there, decisions and deals will be forced to be transparent and accountable."

By providing new avenues to quickly craft hacks that used to require major investment, smart technology will blur the day-to-day management and the long-term planning of cities. Instead of building a new bridge, you might use a model calibrated by high-resolution sensor readings to rejigger signals and tolls to smooth out the flow of traffic. The ability to reprogram instead of rebuild, and evaluate the results immediately through sensors, will allow more experimentation with "soft fixes" and iterative design. It's easy to imagine new cities and neighborhoods where infrastructure and activities are moved around months or even years after they are initially placed, in response to observed patterns of use. Smart technologies could accelerate the growing array of tactical urban interventions and pop-up installations—from food trucks and temporary parks to technology incubators and farmers' markets built inside shipping containers. Much like Cedric Price's *Generator,* the ability to redesign the city on the fly will challenge architects and urban designers to come up with more flexible structures.

At the same time, however, real-time data will be used by citizens to make chronic problems more visible, creating new pressure for long-term fixes. I'm thinking here of dashboard visualizations that are being built on top of real-time open transit data such as "How Fucked is the Orange Line?" which provides up-to-the-minute reminders of delays in Boston's transit system, or "How's Business?" which presents charts of four economic indicators for the city of Chicago (new business licenses, unemployment, building permits for new construction, and foreclosures) alongside a colored summary label—green for "turning around," red for "not looking good," orange for "been better."

Smart technology will also encourage people to engage in local planning debates by highlighting big-picture issues. Neighborhood dashboards that provide ambient information on public displays placed in local shops could visualize larger patterns of change and how they relate to upcoming decisions, much as the Boston transit and Chicago economy examples do. Is there a pattern of gentrification *on this block* visible in recent building permits? How will a proposed project impact traffic, and what does that mean for pedestrian safety *on this corner*? Or you might receive a pop-up message *as you walk past* a proposed redevelopment site, prompting you to weigh in on the latest plans.

Public planning organizations must change profoundly to effectively marry the real-time with the long-term and close the gap on participatory planning. Frank Hebbert works for Open Plans, an advocacy and consulting group that develops open-source technology for cities. When New York City launched its bike-sharing program in 2011, Hebbert led the development of a Web app that allowed citizens to suggest locations for station sites. The public response was massive, yet a lack of transparency made it unclear if or how the transportation planners considered any of the input.

Still, Hebbert is optimistic. He believes we are witnessing a rapid expansion of "tools that help neighborhoods be more prepared when formal planning starts."[44] This could create a virtuous circle, as citizen groups scrutinize open city datasets for warning signs. He speculates, for instance, that analyzing building demolition permits would offer a new holistic and real-time perspective on real estate maneuvers at the block level. The impacts of these private dealings on the community could be better addressed before the fact than afterward.

The days when machines plan our cities are way off. However rapidly they can simulate a new future, humans will remain the key decision makers, and choices about the future of cities will always be disputed. For Joroff, "Strategy will always require a political process to continuously shape what is wanted and what is to be achieved. Both strategy-making and operations require conscious decisions and actions. Neither should not be seen as merely algorithm-driven."[45] But cities that don't find a way to leverage smart technology to make the planning process a more continuous kind of design will fall behind the pace of construction. As *Ekumenopolis*, a recent documentary film on Istanbul's building boom reflects, "Everything changes so fast in this city of 15 million that it is impossible to even take a snap-shot for planning. Plans are outdated even as they are being made."[46] Yet, in this very city, the planners are catching up, and using real-time data to do it. In 2012 IBM helped them redesign the entire city's bus routes based on billions of data points harvested from recent mobile phone movements. The goal—to lay out the bus routes to get people closer to where they were actually going.[47]

Crowdsource with Care

In *Democracy in America*, Alexis de Tocqueville marveled at Americans' propensity to solve problems outside the bounds of government. "Americans of all ages, all conditions, and all dispositions constantly form associations," he famously wrote, "to give entertainments, to found seminaries, to build inns, to construct churches, to diffuse books, to send missionaries to the antipodes; in this manner they found hospitals, prisons, and schools. . . . Wherever at the head of some new undertaking you see the government in France, or a man of rank in England, in the United States you will be sure to find an association."[48] Social technologies are but the latest upgrade to this urge that's embedded in the DNA of American democracy.

Crowdsourcing is a way of tapping and directing the inherent sociability of cities. But as powerful as this approach can be, we need to be cautious. While seemingly progressive, crowdsourcing can also open the door for those who would cut the legs out from under government. Where crowdsourced efforts fill gaps left behind by shrinking budgets, the appearance of an inefficient and ineffective public sector will be difficult to avoid. In cities in the developing world, where crowdsourcing offers services governments have never adequately provided, they may allow for a permanent offloading of obligations. Poor communities may not have the luxury of this level of engagement—the day-to-day realities of survival often leave few resources for volunteerism. Taken to its extreme, crowdsourcing is tantamount to the privatization of public services—the rich will provide for themselves and deny services to those outside their enclaves. Unless we are ready to embrace anarchy and institutionalize unequal access to public services, there will be limits to what crowdsourcing can accomplish.

Crowdsourcing with care means limiting its use to areas where government needs to mobilize citizens around efforts where it lacks capacity, and there is broad consensus over desired outcomes. In a sense, it is the architecture of total civic participation in urban regeneration that Patrick Geddes could only dream of. But as much as crowdsourcing can augment capacity, government needs to ensure that critical public services are delivered to everyone and on time. What happens when helping one part of a crowd hurts another, for instance in traffic avoidance? Do you reward one set of users by revealing secret but limited-capacity, clog-free routes around jams? Or do you redirect everyone and cause entirely new jams? And crowds in and of themselves aren't always a resource—they can be a nuisance too. In 1932, the Regional Plan Association of New York published a pamphlet promoting the need for good city planning. "Some Crowds Are Good," a headline for one section proclaims, illustrated by a parade. But on the next page the image of an overcrowded subway reminds us that "Some Crowds Are Bad."[49] It's a warning we shouldn't forget.

Connect Everyone

Even the most sophisticated crowdsourcing strategy will be undermined if it doesn't engage the right people. But even some of the simplest kinds of smart systems fail to connect everyone.

The consequences of disconnection go beyond just a lack of access. Connection is the means by which people will participate in civic life, not just actively but passively as well. In chapter 6, we saw the inequities in 311 use by non-English speakers in New York and Vancouver, and it's likely this pattern is universal. What's more troubling, however, is that cities increasingly view the data collected by 311 systems as a kind of urban dashboard and early warning system.[50] Aside from inequality of service based on responses to specific complaints, cities may over the long term reallocate resources to trouble hot spots identified by patterns in 311 calls. Given that the most at-risk communities seem to use 311 less, this could produce deep inequities in how public services are provisioned. That 311, arguably the most ubiquitous and simple smart system, brings with it such insidious side effects is a disturbing warning sign. More sophisticated systems of smart governance may have unintended consequences that are even harder to see.

The broader challenge to inclusion in smart cities, however, is that by design everyone is left out. Nothing works until they connect, register, and log in—and any Web start-up trying to build a user base will tell you this is a tricky process to streamline. It's an odd twist for determining eligibility for public services, almost like showing your driver's license to enter a park or queue up at a soup kitchen. Schemes like India's Unique Identification Authority, which will use biometric data to create a digital identity for all 1.2 billion citizens, offer a middle ground. You'll log in with your body, the most minimal of barriers almost everyone will be able to cross. And in addition to reducing barriers to services, it hopes to cut corruption and graft that directly harms the poor and will create an audit trail for the distribution of money and resources. Of course, this is an extreme approach, and it raises enormous concerns around individual privacy.

A special set of issues surrounds how governments connect with the network of nongovernmental organizations (NGOs) that actually monitor and intervene on behalf of the poor and excluded day-to-day. This social sector either supplements or in some cases actually delivers government aid. From 2007 to 2009 I served on Mayor Michael Bloomberg's Broadband Advisory Committee in New York. The group, formed to identify gaps in the city's digital infrastructure and services, held local hearings in communities around the city. In hearing after hearing, nonprofit managers would step up to the microphone and lament their lack of Internet access. Not only were they not engaged in smart-city projects and missing out on the benefits of open government data, they could barely keep up with city government's own electronic reporting requirements for the grants that kept them afloat. Cities need to help foster the development of "data intermediaries" who can provide skills and training needed to make sense of its digital ecosystem.[51] Otherwise, the balance of analytical power between community and commercial interests could be further skewed.

We need to build a systematic evaluation of social sustainability into the planning of new smart-city services. Once we have a sense of the risks, mitigating measures can be designed. In most democracies today (though only in a handful of places in America), there are regulations in place to ensure that plans for new housing, roads, and parks explicitly address the most vulnerable members of society. Technology projects in smart cities must be held accountable to the same standard.

Do Sound Urban Science

We have seen how the introduction of new scientific ideas about cities and data-driven approaches to urban management and planning often bring unwelcome baggage and unintended negative consequences. As I set out to write this book in 2010, a coterie of "hard" scientists—physicists and mathematicians—at the prestigious Santa Fe Institute proclaimed the launch of a new science of cities from their desert retreat. That December, a cover story for the *New York Times Magazine* breathlessly reported on empirical studies of urban growth conducted by Geoffrey West and his colleague Luis Bettencourt. (Ominously, perhaps, the article was written by Jonah Lehrer, who would later resign from his position as staff writer at the *New Yorker* in 2012 following accusation of plagiarism for several articles—not including this one). Homing in on the grandiloquent West as the new champion of rational study of the city, the headline boldly pronounced "A Physicist Solves the City." Despite Lehrer's disparaging claims that "West considers urban theory to be a field without principles, comparing it to physics before Kepler pioneered the laws of planetary motion in the 17th century," and despite any obvious implications for actual policy making or planning, theirs was a significant and welcome addition to the field of urban studies.[52] The big breakthrough, gleaned from data on income, infrastructure, and patents for new innovations, was that as cities grew, they became more productive. A city of 2 million didn't just have twice as much earnings and patents as one with only one million residents, it had double plus 15 percent—a divine gratuity! And that held not just for the good, but the bad as well. Crime and HIV infections were also subject to this superlinear scaling. The process worked in reverse too. Tell West the size of a city, and he could predict its key characteristics. West dazzled audiences around the world with these seemingly universal truths. Yet as my writing came to an end late in 2012, these claims had begun to come under intense scrutiny.

The first salvo came from one of West's and Bettencourt's own colleagues, Carnegie Mellon University statistician Cosma Shalizi, who is himself listed as "external professor" on the Santa Fe Institute website. Shalizi tried to replicate West's and Bettencourt's analysis, and what he discovered was disconcerting for those who had bought into West's elegant theory. In a paper posted to the electronic prepress archive *arXiv*, Shalizi argued that West and Bettencourt had only looked at city-wide figures and not per capita values. "The impressive appearance of scaling displayed," he wrote, "is largely an aggregation artifact, arising from looking at extensive (city-wide) variables rather than intensive (per-capita) ones."[53] Michael Batty, the urban-simulation expert, says that while the scaling effect is still detectable when one converts extensive variables to intensive ones (simply by normalizing, or dividing by population), it is much noisier, or less clear. In general, that is an expected and not immediately disconcerting effect.[54] But what Shalizi also showed was that other explanations could fit the scaling data just as well as the model used by the Santa Fe team. He constructed his own model based on conventional, century-old notions from economic geography that explain why highly productive, specialized businesses tend to cluster in cities. Controlling for just four such industries, he found, "'screens off' the effects of city size on per-capita production." He continued, "there is a weak tendency for per-capita output and income to rise with population, though the relationship is simply too loose to qualify as a scaling law. . . . Qualitatively, this is what one would expect from well-established findings of economic geography."

While Shalizi's paper was ultimately rejected for publication in the *Proceedings of the National Academy of Sciences* (for unknown reasons, as the peer reviews are not made public), the universality of superlinear urban scaling is being called into question by at least one other study. Elsa Arcaute, a researcher at Batty's group in London, has attempted to replicate the results using ward-level data for England and Wales, a much richer level of detail than West and Bettencourt, who worked with units of entire metropolitan areas. What she found is that super-linear scaling appears to occur for some variables, but only if one limits the definition of a city to its dense core. Expand the analysis to include outlying areas of a region and the scaling relationship breaks down. Batty points out that superlinear scaling is also subject to the way different indicators are measured differently in each country, and the distorting effects of policy on land use and migration patterns.[55] The United Kingdom, for instance, has long actively sought to decentralize population and growth from London, which may be one reason why scaling is less evident there. And cities in Europe tend to run into one another, whereas in the United States (where the data fit the Santa Fe model best), there are wide-open spaces separating them. So while superlinear scaling in cities can be found in some places, it clearly isn't as universal as West has argued. The only universal thing about urban scaling may be just how easily it yields to our interventions. "[T]he elegant hypothesis of power-law scaling marked a step forward in our understanding of cities," Shalizi concludes. "But it is now time to leave it behind."[56] Urban scaling isn't quite cold fusion, but it doesn't seem to be the quantum theory of cities either.

This is an important cautionary tale, for the convergence of urbanization and ubiquity will drive demand for rigorous empirical research on cities. In 2012, in New York City alone, three new university departments were established—at Columbia University, New York University/Polytech, and Cornell University—with an explicit focus on applied urban science. These groups, along with others recently launched in London, Chicago, Zurich, and Singapore, will mine the blooming data exhaust of smart cities and deploy new sensory instruments. They will each become what the physicist who leads NYU's effort, Steve Koonin, calls an "urban observatory"—latter-day Outlook Towers where researchers build vast new mirror worlds in search of Gelernter's elusive topsight.[57] The scale and complexity of cities is drawing in bright minds from physics, mathematics, and computer science, just as it intrigued West. But what Shalizi's alternative explanation and Arcaute's detailed geographical analysis tell us is that the old theories are at least as good at explaining what's going on in cities as the new ones. If this new urban science dismisses what has come before it, and fails to ground itself in what has already been discovered, it runs the risk of being at best wrong, and at worst—as it seems West's claims have become—deeply misleading.

As much as West's assertions may have filled our heads with certainty about unconfirmed notions on the nature of urbanization, the collateral damage of these fables so far is probably inconsequential.

Because, in the end, they weren't of much practical use. Intellectually, the idea that cities become more efficient and productive as they grow was fascinating. But what did it mean in terms of policy? That growth was the only sound option? That flew in the face of fifty years of rather sound city-planning practice that sought to manage growth and curb the excesses of unchecked expansion (albeit not always successfully or without unintended consequences). And basic questions about the work's implications remained. How did the process play out? How big could or should cities get? West didn't have any answers for that either. "It's totally unclear if there is a maximum size for cities," he told an audience in New York in 2011.[58] It all seems so disconnected with what has become obvious—trying to grow our way out of ecological collapse is a risky gamble. Constraint, which is what most efforts to promote sustainability really mean, isn't the solution either. The planning vanguard is now embracing the reality of severe climate shocks, and is trying to develop ways of making cities more resilient and able to absorb them. Adaptation, not growth, seems to be how we'll get through the twenty-first century.

A new science of cities is clearly in the making. In fact, it is perhaps the real promise of smart cities. Even if they fail to deliver efficiency, security, sociability, resilience, and transparency—the ambitions of all those stakeholders this book has covered—they will undoubtedly be incredible laboratories for studying how cities grow, adapt, and decline.

"It is of great urgency that we understand cities in a profound and predictive fashion," West has said.[59] His alarm is appropriate. But is it a psychohistorian's dream to think we could compute with any certainty the behavior of something as complex as an entire city, and do it in a way that people can actually use it to solve problems? The field certainly has its work set out for it, and we've seen the many failed attempts to do so. "Data enthusiasm," as Peter Hirshberg called it, rules the day and is fueling the new scientific interest in cities.[60] But even the biggest urban datasets are likely to prove tantalizingly incomplete. As Batty told me during a 2010 interview, "A lot of the old questions which you'd think might be informed by new data are not." When we spoke, he was poring over a new dataset of transactions from the London Underground's Oyster payment-card system. The only problem, he pointed out, was that while some 6.2 million Londoners swiped into the system on an average weekday, only 5.4 million swiped out. Over eight hundred thousand people—nearly 13 percent—"leaked" each day through the sensor web, through exit gates left open during rush hour. "It's as hard as it ever was to get transportation data that is useful," he lamented, "You still need household surveys to actually find out where people are going." A more sound urban science then, will have to ask questions that produce knowledge we can act on, as well as generate data that can seed new theories—it can't just mine data exhaust. As Batty concluded, "There's all this new stuff, but the old questions are still here and they've not been answered."[61]

Slow Data

The big difference between the control revolution that occurred in the cities of the late nineteenth century and the one that's happening now is that the problem then was a lack of communications and a lack of data. Our ability to manufacture and mobilize the physical world outstripped our abilities to communicate and coordinate. Today, the problem is the opposite: we have abundant data and instantaneous communication, and a growing ability not just to sense what is happening but to anticipate and predict what will happen in the future. The problem today isn't figuring out how to accelerate the flow of people, materials, and goods, but rather to try to use less energy by slowing them down. Big data harvested from the exhaust of new sensor networks and everyday transactions promises to shed light on what makes cities tick, streamline their day-to-day management, and inform our long-term plans. But we cannot pretend that we have all the data we need, or that there is always inherent value in mining it. In 1967, as IBM's sales of mainframe computers to corporations and governments were booming, William Bruce Cameron, an American sociologist, made a subtle but stunning observation about the nature of data and society. "It would be nice if all of the data which sociologists require could be enumerated," he wrote, "because then we could run them through IBM machines and draw charts as the economists do. However, not everything that can be counted counts, and not everything that counts can be counted."[62]

For all of our big data, there is still a small universe of crucial bits missing. I think of them as "slow data." Slow data isn't just about plugging the gaps in our sensory infrastructure that prevent researchers like Batty from charting a complete empirical view of the city. It is a tool for unraveling the inevitable spiral of efficiency and consumption that our present conceptions of smart cities could unleash.

The fundamental pitch of technology giants' smart cities is that we can have our cake and eat it too. We can accelerate the flow of information to reduce the flow of resources. But this thinking is flawed. Gains in efficiency often lead to "rebound" consumption. The initial effect of any widely adopted new technology that is more efficient at using a resource—say electricity—is to reduce the cost of that resource as demand falls. But by reducing the cost of a resource, we are spurred to consume more of it, often in other new applications for which it was previously too costly to use as an input. Urban planners have long been familiar with their own version of the rebound effect (or Jevons paradox as it is also known) in transportation planning. Building more roads never reduces traffic for long, but rather unleashes latent demand that was there all along. When congestion is reduced due to the new capacity, the opportunity cost of driving falls, spurring drivers who would never have ventured onto the previously clogged road to sally forth.

Over the coming decades, we'll witness just such a process play out as automated vehicles take to the road. So far, the excitement over innovations like Google's self-driving car has been about safety and convenience. You'll be able to surf the net during your commute. You'll never have to worry about your drunken teenager wrapping the family sedan around a telephone pole. But the even greater economic potential of self-driving cars is that they could potentially double road capacity by reducing spacing between cars and jams caused by a whole host of idiosyncratic human behaviors. If that spurs people who would have stayed home to take new trips, we'll have to double fuel economy just to hold even. Reducing overall emissions would require dramatic increases in efficiency to keep up with the expanding volume of traffic.

It shouldn't surprise us to find these cycles of increasing consumption that lead nowhere. They are endemic to industrial capitalism. In *The Jungle*, Upton Sinclair's reality drama about the harsh working conditions of the Chicago stockyards at the turn of the twentieth century, we learn about the process of "speeding-up the gang" used by slaughterhouse bosses to boost output. "There were portions of the work which determined the pace of the rest, and for these they had picked men whom they paid high wages, and whom they changed frequently. You might easily pick out these pacemakers, for they worked under the eye of the bosses, and they worked like men possessed."[63] In smart cities, technologies of automation take the place of the speed-up men. They may whisk away the consequences of consumption, and make us more efficient as individuals at the things we do now. But they do nothing to stack the deck for a lower-emissions civilization in the long run.

By automating conservation, designing it in, these smart cities don't offer us any incentives to *decide* to cut back. That's where slow data comes in. Slow data must be collected, sparingly and by design, not harvested opportunistically from data exhaust. Rather than hide the trade-offs between consumption and conservation, slow data makes it explicit. It makes us choose. And slow data leverages our humanness, by generating social interactions that help address these vexing problems.

As an example, take the problem of finding lost objects. The big data approach would be to tag and track everything, perhaps using RFID, the wireless barcode technology whose tiny plastic tags cost just a few cents apiece. They are already used in clothing stores, where they expedite checkout and reduce the cost of inventory and security. As an array of scanners roll out across the smart city, the Internet of Things will become searchable in real time. All it would take to find anything, anywhere, would be a piece of software to scan the logs—trillions of measurements which, if collected in one place, will be the biggest set of big data there is.

What if, instead, we just helped each other find things? Instead of creating an infrastructure for machine surveillance to find our lost stuff, we could build one for social cooperation that provides the same capability—but faster, cheaper, and with positive social side effects. This is the idea behind PhoundIt, an app that bills itself as "lost and found, redesigned for the connected city." Using the Foursquare API, PhoundIt takes reports of lost items, and alerts users when they check in somewhere to be on the lookout for them. When something is found, there are tools to arrange a safe return. As Elan Miller, the project's founder explains, the goal is to "make it easy for the community to act on their inherent goodwill and inspire others to do the same."[64] Phoundit demands a lot from us, but unlike the automated system, it dangles the enticing prospect of meaningful human contact. It appeals to basic human altruism, but also our inherent desire to be social and seek out new relationships. There is a sustainability angle too: instead of consuming more by simply replacing lost objects, PhoundIt's users extend their working life. There's also no need to manufacture billions of RFID tags and a global infrastructure to track them.

The lesson is: don't lose sight of the slow data in the torrent of big data. The real opportunity to design killer apps for smart cities lies in those niches where a couple of heavily value-laded bits can be created—just as the Foursquare check-in and the Facebook "like" have. Slow data's power is its ability to induce behavior change—as we saw with the Botanicalls project that matched a tweeting houseplant with a caregiving network of grad students. And slow data can complement big data—whenever efficiency is warranted, it should be paired with mechanisms that deliver those behavior-changing bits into the foreground of our social lives, where we can think about the trade-offs. Big data may streamline our wasteful ways, but it will take slow data to change them. Big data may make us lean and mean. Slow data will speak to our souls.

I'm often asked, "What's the smartest city?"

My answer is always the same. "The one you live in."

It sounds glib, but I'm serious. The idea of a single, utopian design for the smart city has kept us from the hard work of building a rich and varied collection of ones that we can actually live with. Since 2008, the vision of our urban future has come to be dominated by companies that would repeat the cookie-cutter city designs of the twentieth century on a planetary scale, powered by the technology of global enterprise. Our mayors are putting their own spin on these designs, but they can't solve all of our problems.

The answer lies at the grass roots. I see it blossoming everywhere as we take these tools out into the streets and use them to reimagine and remake our world. We thought the Internet was about transcending the globe, and then it took a hyperlocal turn and became about swapping reviews of restaurants and getting free coupons for the local shop. We thought it would isolate social groups, and then it connected us all into one big network. We thought it was about staying home and looking at physics papers or LOLcats, and then in just a few years it powered over a million meatspace meetups.

Smart-city hackers can't do it alone. While we can show our business leaders and politicians how to build a more just, social, and sustainable future, we need their help to reach critical mass. Like Patrick Geddes, I believe that it will take a social movement that enlists science, the humanities, and us all to address the challenges we face building a planet of cities that can survive. Whether we call it an urban operating system or the industrial Internet, something really big is booting up in the half-million-plus civic laboratories on planet Earth.

Are you going to help build it?

You have everything you need.

Epilogue

The technology of the smart city is well hidden. It lives underground and inside walls, hides in our pockets, and rides on the airwaves around us. But when a jackhammer-wielding crew peels back the hard asphalt of a city street, the guts of the metropolis are revealed. Among sewer pipes and gas lines laid many years ago, a digital nervous system now resides.

Recently, one of these infrastructural incisions opened up not far from where I first discovered Dennis Crowley's Dodgeball app a decade ago. Back then, smartphones were still years away. But the area around Manhattan's Astor Place was now a hub for the business of big data—much of it generated by apps like Facebook's—running on millions of those little supercomputers in our pockets. The social-networking giant had recently relocated its New York staff of more than three hundred to the old Wanamaker's Department Store building on the west side of the busy square.

But Facebook wasn't to blame for the new subterranean action. An ever bigger data hog was in the neighborhood. Months earlier, IBM had announced that it would locate its new Watson Group in One Astor Plaza, just across the street from Facebook. Watson was the most ambitious effort in history to turn artificial intelligence into big bucks, a collection of algorithms in the cloud designed to find meaning in big data. Watson promised to bring superintelligence to all manner of human decision making, from the analysis of CAT scans to the deployment of police forces.

IBM's $1 billion investment would put two thousand geeks in Crowley's old stomping grounds. The sleek, jet-black tower loomed ominously over the throngs of students and skate punks that clustered around its base. Amid the city buses and food trucks that jostled for curb space stood a gray van, parked aside a freshly dug trench in the street. "Empire City Subway" read the lettering on its side. But ECS (as the company is known) doesn't build subway tunnels for trains. It builds tubes for telecommunications.

The smart city had never seemed more concrete. Watson needed data, and lots of it, to feed its pattern-matching prowess. The digital plumbers from ECS were tearing the street apart to make way for the fiber-optic cables that would carry bits to IBM's machine minds.

Smart Cities was an attempt to chalk out a big, forward-looking picture of what digital technology means for cities.

But one of the challenges of forecasting is that while you may get the general direction of change right, nailing the magnitude can be tricky. If you try to stay plausible by today's assumptions, you run the risk of underestimating the potential for sudden, rapid change in the future. With smart cities, trends that only recently appeared small on the horizon now loom larger and larger. Everything seems to be speeding up, getting bigger, or getting worse than was expected.

The title of Chapter 1 was chosen deliberately to shock and awe. Borrowing a figure from a key market forecast, "The $100 Billion Jackpot" was the pie big technology corporations hoped to carve up among themselves by 2020. Yet just a few weeks after this book's first publication, and barely three years after that earlier estimate, the UK government announced its own prediction—over $650 billion annually by 2020. Tabulated by analysts at engineering giant Arup, this new calculation assumed a much broader definition of the industry. But it corroborated other intelligence. The market was vaster than anyone had dared imagine.

Yet even as the financial outlook for smart cities went supernova, fears that the biggest smart-city outfitters would dominate the action turned out to be overblown. In this case, the trend line fell far short of what many expected. None of the big technology and engineering companies that sought early on to corner the smart cities market—IBM, Cisco, and Siemens—have met their revenue targets. A key goal of *Smart Cities* was to get people thinking about the risks of letting big tech companies design future cities. It now seems these companies overestimated their ability to take the high ground.

In the meantime, governments and civic hackers swooped in with a surprisingly ample war chest. In 2013, the UK government committed £50 million (about $84 million) over five years to a smart-city tech accelerator in London, the Future Cities Catapult Centre. Meanwhile, the Chinese government unveiled a crash program to build a hundred smart cities, towns, and districts. The United Kingdom hopes to kick-start businesses that can export technology and solutions globally. China is looking to address its massive domestic market, which it pegs to grow to some $80 billion a year by 2025. According to an assessment published by the Knight Foundation, more than $430 million has been invested in over two hundred civic tech start-ups since 2011. Over $1 billion of venture investments targeted Internet of Things start-ups in 2013 alone.

Unfortunately, the risks inherent in smart cities have also snowballed. The Heartbleed security bug exposed in April 2014 showed how sloppy programming can introduce catastrophic flaws into the infrastructure of smart cities. In this case, a late-night coding error by a volunteer programmer went unnoticed for over two years by the many eyeballs of the open-source community, creating a pervasive hole in the most important tool for securing the Internet. Just before Thanksgiving in 2013, one of the largest thefts of sensitive consumer data in history occurred when hackers infiltrated the megaretailer

Target's corporate intranet through an Internet-connected heating-and-cooling system. Similarly, reminders of the brittleness of smart-city systems are a daily occurrence—for instance, the February 2014 breakdown of taxi-hailing app Uber after a California data-center outage. And perhaps no level of alarm about government surveillance of citizens' communications is unjustified in light of National Security Agency whistle-blower Edward Snowden's alarming revelations during the summer of 2013.

Taken together, these rapid accelerations of the forces at play in smart cities foretell a colossal and potentially chaotic period of growth and transformation. But amid the haze of uncertainty, one final expectation-resetting investment boom provides hope: the birth of dozens of university research centers focused on the study of cities. For nearly a decade, MIT's SENSEable City Lab seemed to stand alone in its focused exploration of big cities through the massive and growing plume of exhaust data they generate. But now, even more ambitious efforts to build a new quantitative understanding of cities are spinning up in New York, London, Chicago, Zurich, Singapore, Amsterdam, Madrid, Glasgow, and elsewhere. Collectively, they represent the greatest infusion of funding, talent, and new ideas into urban studies in a century.

The timing couldn't be better. Because the one trend that seems to be bucking my expectation-beating pattern is the pace of urbanization itself. Even in China, India, and Africa, the expansion of cities seems to have abated, if ever so slightly. So while cities will continue to evolve on a scale and level of complexity we've never seen, they may yet grant us enough time to see if this crash program in decoding the nature of urbanization will yield fruit. It is becoming clear that some big discoveries are tantalizingly near. If we exploit these sorts of lessons and redesign our blueprints for the last great era of city building that lies ahead, we might just make it to the other side of the coming century.

Acknowledgments

I've had the great fortune of working with a number of mentors who have shaped my understanding of cities and technology and how they shape each other. More than anyone, Mitchell Moss at New York University taught me firsthand what cities are and how they work, and provided the opportunity for me to explore the emerging urban geography of the Internet in the mid-1990s. The late William Mitchell at MIT inspired and encouraged me to think more deeply about the role of place and physical design in the smart city. Our discussions about wireless networks and digital resilience in New York after September 11 frame many of the ideas in this book, and I proudly borrow its title from the research group he started at the Media Lab in 2003. Frank Popper at Rutgers University closed the loop between my interest in cities and my fascination with computers in 1995, handing me a copy of the US Office of Technology Assessment's report *The Technological Reshaping of Metropolitan America*. Over the last decade, Michael Joroff and Dennis Frenchman at MIT pulled me in on smart-city design projects all over the world and patiently demystified the players and strategies at work in the modern city-building industry.

The Institute for the Future is the world's leading center for long-term thinking and has been my intellectual home since 2005. Without the support of my colleagues, this book would never have come to be. Marina Gorbis and Bob Johansen have helped me understand the science and art of long-range forecasting. The overarching framework for this book—the conflict between industry and the grassroots visions for smart cities—grew out of work with Kathi Vian and Michael Liebhold in 2006 on the future of context-aware computing. Kim Lawrence was indispensable in helping arrange the leave of absence that gave me the time to write.

In New York City I've been surrounded by a coterie of fellow thinkers and doers pushing the frontier of smart cities forward, and our conversations helped make this a better book—especially Greg Lindsay, Adam Greenfield, Laura Forlano, Andrew Blum, Jake Barton, Frank Hebbert, and Hugh O'Neill. Others outside the Big Apple who offered helpful comments on early drafts include Anna Ponting, Alex Soojung-Kim Pang, Francisca Rojas, and Rob Goodspeed. From my NYCwireless compatriots Terry Schmidt, Dustin Goodwin, Joe Plotkin, Dana Spiegel, Ben Serebin and Jacob Farkas I learned firsthand how to hack smart cities together at the hardware level.

The keenest eye of all has been that of my editor Brendan Curry, whose course corrections vastly improved this manuscript. My agent, Zoë Pagnamenta, has been an adept guide in the world of publishing. In the course of checking facts, Patricia Chui discovered dozens of details that have greatly enriched the stories told herein. Amanda Alampi compiled the hundreds of notes documenting my research, and, more importantly, helped me understand how to share the stories in this book through social media.

Generous financial support from Benjamin de la Peña of the Rockefeller Foundation through a grant for a study on "the future of cities, information and inclusion" planted the seed for this book, and additional follow-on funding from Carol Coletta of CEOs for Cities supported my initial writing. The Kauffman Foundation supported research on the role of entrepreneurs and start-ups in building smart cities. The Frederick Lewis Allen Memorial Room at the New York

Public Library and the S. C. Williams Library at Stevens Institute of Technology provided workspaces for research and writing.

Richard and Roberta Townsend always allowed me from a young age to choose my own direction in life, and provided constant encouragement and support. My wife Nicole has been there always, as a sounding board, honing my reasoning and helping me actually turn some of these ideas into real projects. Finally, my brothers John Townsend and Bill Townsend, who were my original urban mentors, showed me the wonders of Boston and Washington as a teenager, and spurred my love of the city forever.

Notes

Preface

1 "America's New Mobile Majority: A Look at Smartphone Owners in the U.S." *Nielsen Wire*, blog, last modified May 7, 2012, http://blog.nielsen.com/nielsenwire/?p=31688.

Introduction. Urbanization and Ubiquity

1 *World Urbanization Prospects: The 2007 Revision* (New York: United Nations, Department of Economic and Social Affairs, Population Division, February 2008), 1.
2 *World Urbanization Prospects: The 2009 Revision* (New York: United Nations, Department of Economic and Social Affairs, Population Division, March 2010), 1.
3 Urban population in 1900: "Human Population: Urbanization" (Washington, DC: Population Reference Bureau, 2007), http://www.prb.org/Educators/TeachersGuides/HumanPopulation/Urbanization.aspx; world population in 1900: *The World At Six Billion* (New York: United Nations, Department of Economic and Social Affairs, Population Division, October 1999), 4.
4 *World Urbanization Prospects: The 2011 Revision* (New York: United Nations, Department of Economic and Social Affairs, Population Division, March 2012), 1.

5 Author's calculation based on global population forecast in *World Population Prospects: The 2010 Revision* (New York: United Nations, Department of Economic and Social Affairs, Population Division, May 2011), xiii, and urbanization forecast of 70–80 percent in Shlomo Angel, *Planet of Cities* (Cambridge, MA: Lincoln Institute of Land Policy, September 2012).

6 Shirish Sankhe et al., "India's urban awakening: Building inclusive cities, sustaining economic growth" (New York: McKinsey Global Institute, McKinsey & Co., April 2010), http://www.mckinsey.com/insights/mgi/research/urbanization/urban_awakening_in_india.

7 "Twenty New Cities to Be Set Up in China Every Year," *People's Daily*, last modified August 14, 2000, http://english.people.com.cn/english/200008/14/eng20000814_48177.html.

8 Slum population: *State of the World's Cities 2012/2013: Prosperity of Cities, World Urban Forum Edition* (Nairobi, Kenya: UN-HABITAT, 2012), 100; population projection, lecture, Joan Clos, Director, UN-HABITAT, Smart Cities Expo 2011, Barcelona, Spain, November 29, 2011.

9 D. Kissick et al., *Housing for All: Essential for Economic, Social, and Civic Development*, manuscript prepared for the World Urban Forum III by PADCO/AECOM, 2006, http://www.hrc.co.nz/wp-content/uploads/2012/10/housing_for_all.pdf, 1.

10 "Key Global Telecom Indicators for the World Telecommunication Service Sector," *International Telecommunication Union*, last modified November 16, 2011, http://www.itu.int/ITU-D/ict/statistics/at_glance/KeyTelecom.html.

11 "Key Global Telecom Indicators," *International Telecommunication Union*.

12 Mary Meeker, "KCBP Internet Trends," presentation, D10 Conference, Rancho Palos Verdes, CA, May 30, 2012, http://www.scribd.com/doc/95259089/KPCB-Internet-Trends-2012.

13 Ted Schadler and John C. McCarthy, "Mobile is the New Face of Engagement" (Cambridge, MA: Forrester Research, Inc., February 13, 2012), http://www.forrester.com/Mobile+Is+The+New+Face+Of+Engagement/fulltext/-/E-RES60544?objectid=RES60544.

14 "U.S. Wireless Quick Facts," Cellular Telecommunications Industry Association, n.d., accessed February 3, 2013, http://www.ctia.org/consumer_info/index.cfm/AID/10323.

15 Massoud Amin, "North American Electricity Infrastructure: System Security, Quality, Reliability, Availability, and Efficiency Challenges and their Societal Impacts," in *Continuing Crises in National Transmission Infrastructure: Impacts and Options for Modernization*, National Science Foundation (NSF),

June 2004, 1; *CTIA Semi-Annual Wireless Industry Survey* (Washington, DC: Cellular Telecommunications Industry Association, 2012), http://files.ctia.org/pdf/CTIA_Survey_MY_2012_Graphics-_final.pdf.

16 Dave Evans, "The Internet of Things: How the Next Evolution of the Internet is Changing Everything," (San Jose, CA: Cisco Systems, April 2011), http://www.cisco.com/web/about/ac79/docs/innov/IoT_IBSG_0411FINAL.pdf, 3.

17 Evans, "The Internet of Things," 3.

18 "Cisco Visual Networking Index: Global Mobile Data Traffic Forecast Update, 2011–2016," February 14, 2012, http://www.cisco.com/en/US/solutions/collateral/ns341/ns525/ns537/ns705/ns827/white_paper_c11-520862.html.

19 "How Much Is A Petabyte?" *Mozy*, blog, last modified July 2, 2009, http://mozy.com/blog,/misc/how-much-is-a-petabyte/.

20 Gary Locke, US ambassador to China, interview by Charlie Rose, January 16, 2012, http://www.charlierose.com/view/interview/12091.

21 "Charles Minot," National Railroad Hall of Fame: Galesburg, IL, n.d., accessed October 17, 2012, http://www.nrrhof.org/pages/minot.php; Henry D. Estabrook, "The First Train Order by Telegraph," *B&O Magazine: Baltimore and Ohio Employees Magazine*, July 1913, 27.

22 Joel A. Tarr with T. S. Finholt and D. Goodman, "The City and the Telegraph: Urban Telecommunications in the Pre-Telephone Era," *Journal of Urban History* 14 (1987): 38–80, reprinted in Stephen Graham (ed.), *The Cybercities Reader* (London: Routledge, 2003).

23 Herbert Casson, *The History of the Telephone* (Chicago: A. C. McClurg, 1910), 222.

24 "The Knowledge Explosion," BBC Horizon series, originally broadcast September 21, 1964, archived at http://www.youtube.com/watch?v=KT_8-pjuctM.

25 "City vs. Country: Tom Peters & George Gilder debate the impact of technology on location," *Forbes ASAP*, February 27, 1995, http://business.highbeam.com/392705/article-1G1-16514107/city-vs-country-tom-peters-george-gilder-debate-impact.

26 David McCandless, "Financial Times Graphic World," display at Grand Central Station, New York, March 27–29, 2012.

27 Robert Caro, *The Power Broker: Robert Moses and the Fall of New York* (New York: Vintage, 1975), 849.

28 Caro, *The Power Broker*, 508.

29 "Global Investment in Smart City Technology Infrastructure to Total $108 Billion by 2020," *Pike Research*, last modified May 23, 2011, http://www.pike

research.com/newsroom/global-investment-in-smart-city-technology-infra structure-to-total-108-billion-by-2020.

30 Daniel Fisher, "Urban Outfitter," *Forbes*, May 9, 2011, 92.

31 Sascha Haselmeyer, lecture, INTA33 World Urban Development Congress, Kaoshiung, Taiwan, October 5, 2009.

32 "The Explosive Growth of Bus Rapid Transit," The Dirt, blog, *American Society of Landscape Architects*, last modified January 27, 2011, http://dirt.asla.org/2011/01/27/the-explosive-growth-of-bus-rapid-transit/.

33 Peter Jamison, "BART Jams Cell Phone Service to Shut Down Protests," *SF Weekly:* The Snitch, blog, August 12, 201, http://blogs.sfweekly.com/thesnitch/2011/08/bart_cell_phones.php; BlackBerry: Josh Halliday, "David Cameron considers banning suspected rioters from social media," *The Guardian*, August 11, 2011, http://www.guardian.co.uk/media/2011/aug/11/david-cameron-rioters-social-media; social media: Chris Hogg, "In wake of London riots, UK considers social media bans," *Future of Media*, blog, http://www.futureofmediaevents.com/2011/08/11/in-wake-of-london-riots-uk-considers-social-media-bans/#ixzz24xS7KHKP.

34 Solomon Benjamin et al., "Bhoomi: 'E-Governance,' Or, An Anti-Politics Machine Necessary to Globalize Bangalore?" CASUM-m, Bangalore, India, January 2007, http://casumm.files.wordpress.com/2008/09/bhoomi-e-governance.pdf.

35 Kevin Donovan, "Seeing Like a Slum: Towards Open, Deliberative Development," *Georgetown Journal of International Affairs* 13, no. 1 (2012): 97.

36 Jeremy Bentham. *The Panopticon Writings* (London: Verso, 1995), 29–95.

37 Farah Mohamed, "Sen. Franken on facial recognition and Facebook," *Planet Washington*, last modified July 18, 2012, http://blogs.mcclatchydc.com/washington/2012/07/sen-franken-on-facial-recognition-adnd-facebook.html.

38 Adam Harvey, *CV Dazzle*, n.d., accessed August 26, 2012, http://cvdazzle.com.

39 Jane Jacobs, *The Death and Life of Great American Cities* (New York: Random House, 1961), 238.

40 Walter Lippmann, *New York Herald Tribune*, June 6, 1939, quoted in Robert W. Rydell, *World of Fairs: The Century-of-Progress Expositions* (Chicago: University of Chicago Press, 1993), 115.

Chapter 1. The $100 Billion Jackpot

1 Henrik Schoenefeldt, "191: The Building of the Great Exhibition of 1851, an Environmental Design Experiment" (Cambridge: The Martin Centre for

Architectural and Urban Studies, University of Cambridge, n.d.), http://kent.academia.edu/HenrikSchoenefeldt/Papers/118104/The_Building_of_the_Great_Exhibition_of_1851_-_an_Environmental_Design_Experiment.

2 Terence Riley, *The Changing of the Avant-Garde: Visionary Architectural Drawings from the Howard Gilman Collection* (New York: Museum of Modern Art, 2002), 150.

3 "A Walking City," Archigram Archival Project, Project for Experimental Practice, University of Westminster, 2010, http://archigram.westminster.ac.uk/project.php?id=60.

4 Michael Sorkin, "Amazing Archigram," *Metropolis*, April 1998, http://www.metropolismag.com/html/content_0498/ap98what.htm.

5 Riley, *The Changing of the Avant-Garde.*

6 Molly Wright Steenson, "Cedric Price's Generator," *Crit* 69 (2010), 14.

7 Steenson, "Cedric Price's Generator," 15.

8 Robert Lenzer and Tomas Kellner, "Fall of the House of Gilman," *Forbes*, last modified August 11, 2003, http://www.forbes.com/forbes/2003/0811/068.html.

9 Royston Landau, "Cedric Price," Museum of Modern Art, last modified 2009, http://www.moma.org/collection/artist.php?artist_id=7986.

10 Paul Ehrlich and Ira Goldschmidt, "Building Automation: Green Intelligent Buildings—A Brief History," *Engineered Systems*, March 1, 2008, http://www.esmagazine.com/Articles/Column/BNP_GUID_9-5-2006_A_10000000000000271363.

11 John D. Kasarda and Greg Lindsay, *Aerotropolis: The Way We'll Live Next* (New York: Farrar, Straus and Giroux, 2011), 357.

12 "RFID/USN Cluster to Be Built in Songdo By 2010," *Korea IT Times*, last modified October 31, 2005. http://www.koreaittimes.com/story/2162/rfidusn-cluster-be-built-songdo-2010.

13 Charles Arthur, "This City Will Change the World," *BBC Knowledge*, May/June 2012, 28.

14 Charles Arthur, "The Thinking City," *BBC Focus*, January 2012, 55–59.

15 Kasarda and Lindsay, *Aerotropolis*, 353.

16 John Boudreau, "Cisco wires 'city in a box' for fast-growing Asia," *San Jose Mercury News*, last modified June 8, 2010, http://www.newsobserver.com/2010/06/08/520176/cisco-wires-city-in-a-box-for.html.

17 Seoul Development Institute, *Seoul, 20th Century: Growth and Change of the Last 100 Years* (Seoul: Seoul Development Institute, 2003), 14.

18 Anthony M. Townsend, "Seoul: Birth of A Broadband Metropolis," *Environment and Planning B* 34, no. 3 (2007): 396–413.

19 "Global building automation market predicted to grow 3 percent by 2015," *SustainableBusiness.com News*, last modified February 4, 2010, http://www.sustainablebusiness.com/index.cfm/go/news.display/id/19697.

20 Lewis Mumford, *The City in History: Its Origins, Its Transformations, and Its Prospects* (New York: MJF Books, 1997), 527.

21 Philip Carter, Bill Rojas, and Mayur Sahni, "Delivering Next-Generation Citizen Services: Assessing the Environmental, Social and Economic Impact of Intelligent X on Future Cities and Communities," IDC, June 2011, http://www.cisco.com/web/strategy/docs/scc/whitepaper_cisco_scc_idc.pdf.

22 John Frazer, lecture, Forum on Future Cities, MIT SENSEable City Lab and the Rockefeller Foundation, Cambridge, MA, April 13, 2011, http://techtv.mit.edu/collections/senseable/videos/12305-changing-research.

23 "Why Songdo: Sustainable City," accessed January 24, 2013, http://www.songdo.com/songdo-international-business-district/why-songdo/sustainable-city.aspx.

24 Songdo International Business District "Master Plan," http://www.songdo.com/songdo-international-business-district/the-city/master-plan.aspx and "Living," http://www.songdo.com/songdo-international-business-district/the-city/living.aspx, accessed September 25, 2012.

25 Tim Edelston, "Still Time for Songdo City to Protect Biodiversity," *Korea Times*, last modified January 8, 2012, http://www.koreatimes.co.kr/www/news/opinon/2012/01/137_102458.html.

26 Viren Doshi, Gary Schulman, and Daniel Gabaldon, *strategy + business*, last modified February 28, 2007, http://www.strategy-business.com/article/07104.

27 "Reinventing the City," World Wildlife Fund (WWF), 2010, http://www.wwf.se/source.php/1285816/Reinventing%20the%20City_FINAL_WWF-rapport_2010.pdf, 2.

28 Jonathan D. Miller, "Infrastructure 2011: A Strategic Priority," Urban Land Institute and Ernst & Young, 2011, http://www.uli.org/ResearchAndPublications/%7E/media/Documents/ResearchAndPublications/Reports/Infrastructure/Infrastructure2011.ashx.

29 Ian Marlow, lecture, "X-Cities 4: Cities-as-Service," Columbia University Studio-X, New York, April 19, 2012.

30 "Global Investment in Smart City Technology Infrastructure," *Pike Research*.

31 "Smart City Technologies Will Grow Fivefold to Exceed $39 Billion in 2016," *ABI Research*, last modified July 6, 2011, http://www.abiresearch.com/press/3715-Smart+City+Technologies+Will+Grow+Fivefold+to+Exceed+$39+Billion+in+2016.

32 Colin Harrison, remarks, Ideas Economy: Intelligent Infrastructure, *The Economist*, New York City, February 16, 2011.

33 "Smart Cities: Transforming the 21st century city via the creative use of technology," ARUP, last modified September 1, 2010, http://www.arup.com/Publications/Smart_Cities.aspx.

34 Stephen Graham, "The end of geography or the explosion of place? Conceptualizing space, place and information technology," *Progress in Human Geography* 22, no. 2 (1998): 165–85.

35 "International Energy Outlook 2011," DOE/EIA-0484(2011), U.S. Energy Information Administration, September 19, 2011, http://www.eia.gov/oiaf/ieo/electricity.html.

36 Amin, "North American Electricity Infrastructure: System Security, Quality, Reliability, Availability, and Efficiency Challenges and their Societal Impacts," 1.

37 "Electric Power Annual," U.S. Energy Information Administration, last modified November 9, 2011, http://www.eia.gov/electricity/annual/html/tablees1.cfm.

38 Tim Wu, *The Master Switch: The Rise and Fall of Information Empires* (New York: Knopf, 2010), 102–3.

39 "75% of US Electric Meters to be Smart Meters by 2016," In-Stat press release, March 5, 2012, http://www.instat.com/press.asp?ID=3352&sku=IN1104731WH.

40 "Historical Figures in Telecommunications," International Telecommunications Union, last modified February 11, 2010, http://www.itu.int/en/history/overview/Pages/figures.aspx.

41 Urs Fitze, "No Longer A One-Way Street," *Pictures of the Future*, Spring 2011, 22, http://www.siemens.com/innovation/pool/en/publikationen/publications_pof/pof_spring_2011/pof_0111_strom_smartgrid_en.pdf.

42 Edwin D. Hill, "New Challenges Demand New Solutions: IBEW Leader Charts Energy Future," *EnergyBiz*, September/October 2007, http://energycentral.fileburst.com/EnergyBizOnline/2007-5-sep-oct/Financial_Front_New_Challenges.pdf.

43 Martin Rosenberg, "Continental Grid Vision Needed," *RenewableEnergyWorld.com* blog, last modified December 11, 2007, http://www.renewableenergyworld.com/rea/news/article/2007/12/continental-grid-vision-needed-50777.

44 "Company development 1847–1865," Siemens, n.d., http://www.siemens.com/history/en/history/1847_1865_beginnings_and_initial_expansion.htm.

45 Jeff St. John, "How Siemens is Tackling the Smart Grid," *GigaOM*, last modified June 24, 2010, http://gigaom.com/cleantech/how-siemens-is-tackling-the-smart-grid/.

46 "Siemens CEO Peter Löscher: We're on the threshold of a new electric age," Siemens press release, December 15, 2010, http://www.siemens.com/press/en/pressrelease/?press=/en/pressrelease/2010/corporate_communication/axx20101227.htm.

47 "75% of US Electric Meters to be Smart Meters by 2016," In-Stat press release, March 5, 2012, http://www.fiercetelecom.com/press-releases/75-us-electric-meters-will-be-smart-meters-2016.

48 Chris Nelder, "Why baseload power is doomed," *SmartPlanet*, blog, last modified March 28, 2012, http://www.smartplanet.com/blog/energy-futurist/why-baseload-power-is-doomed/445.

49 Massoud Amin, "North American Electricity Infrastructure: System Security, Quality, Reliability, Availability, and Efficiency Challenges and their Societal Impacts," in *Continuing Crises in National Transmission Infrastructure: Impacts and Options for Modernization*, National Science Foundation (NSF), June 2004.

50 Fitze, "No Longer A One-Way Street," 23.

51 Tim Schröder, "Automation's Ground Floor Opportunity," *Pictures of the Future*, Spring 2011, 19, http://www.siemens.com/innovation/apps/pof_microsite/_pof-spring-2011/_pdf/pof_0111_strom_buildings_en.pdf.

52 Eric Paulos, lecture, "Forum on Future Cities," MIT SENSEable City Lab and the Rockefeller Foundation, Cambridge, MA, April 13, 2011, http://techtv.mit.edu/collections/senseable/videos/12305-changing-research; For a thorough treatment see Eric Paulos and James Pierce, "Citizen Energy: Towards Populist Interactive Micro-Energy Production," n.d., http://www.paulos.net/papers/2011/Citizen_Energy_HICSS2011.pdf.

53 James R. Beniger, *The Control Revolution: Technological and Economic Origins of the Information Society* (Cambridge, MA: Harvard University Press, 1986), 12.

54 Eduardo Aibar and Wiebe E. Bikjer, "Constructing A City: The Cerdá Plan for the Extension of Barcelona," *Science, Technology, & Human Values* 22, no. 1 (1997): 3.

55 Ildefons Cerdà, *Teoría General de la Urbanización* (Madrid: Imprenta Española: 1867), 595, quoted in Arturo Soria y Puig, *Cerda: The Five Bases of the General Theory of Urbanization* (Madrid: Electa, 1999), 57.

56 Salvador Tarragó and Francesc Magrinyà, *Cerdà, Urbs i Territori: Planning Beyond the Urban* (Madrid: Electa, 1996), 202.

57 Tarragó and Magrinyà, 190.

58 Tom Standage, *The Victorian Internet: The Remarkable Story of the Telegraph and the Nineteenth Century's On-line Pioneers* (New York: Berkley Books, 1999).

59 "Cisco Launches Innovation Centre to Build Next Generation Services in Singapore," Cisco Systems press release, December 12, 2008, http://investor.cisco.com/releasedetail.cfm?ReleaseID=354147.

60 "Cisco's Wim Elfrink: 'Today, We Are Seeing What I Call the Globalization of the Corporate Brain,'" *India Knowledge@Wharton*, Wharton School, University of Pennsylvania, last modified July 16, 2009, http://knowledge.wharton.upenn.edu/india/article.cfm?articleid=4395.

61 "Smart + Connected Communities: Changing a Community, a Country a World," Cisco Systems, June 2010, 3, http://www.cisco.com/web/strategy/docs/scc/09CS2326_SCC_BrochureForWest_r3_112409.pdf.

62 "Cisco Visual Networking Index: Global Mobile Data Traffic Forecast Update, 2011–2016," Cisco Systems, last modified February 14, 2012, http://www.cisco.com/en/US/solutions/collateral/ns341/ns525/ns537/ns705/ns827/white_paper_c11-520862.html.

63 "How Virtual Meetings Provide Substantial Business Value and User Benefits," Cisco Systems, n.d., accessed September 25, 2012, http://www.cisco.com/web/about/ciscoitatwork/downloads/ciscoitatwork/pdf/Cisco_IT_Case_Study_TelePresence_Benefits.pdf.

64 Daniel Brook, "The Rise and Fall and Rise of New Shanghai," *Foreign Policy*, September/October 2013, last modified August 13, 2012, http://www.foreignpolicy.com/articles/2012/08/13/the_rise_and_fall_and_rise_of_new_shanghai.

65 "Smart + Connected Life Video," Cisco Systems, n.d., http://www.cisco.com/web/CN/expo/en/pavilion.html.

66 "Smart + Connected Life Video."

67 Eliza Strickland, "Cisco Bets on South Korean Smart City," *IEEE Spectrum*, last modified November 29, 2011, http://spectrum.ieee.org/telecom/internet/cisco-bets-on-south-korean-smart-city.

68 Alex Soojung-Kim Pang, "Mobility, Convergence, and the End of Cyberspace," in Kristof Nyiri, ed., *Towards a Philosophy of Telecommunications Convergence* (Vienna: Passagen Verlag, 2008), 55–62.

69 Matt Novak, "The World's First Carphone," *Paleofuture*, blog, *Smithsonian Magazine*, last modified January 25, 2012, http://blogs.smithsonianmag.com/paleofuture/2012/01/the-worlds-first-carphone/.

70 Novak, "The World's First Carphone."

71 "First FM Portable Two-Way Radio," Motorola Solutions, accessed September 25, 2012, http://www.motorolasolutions.com/US-EN/About/Company+Overview/History/Explore+Motorola+Heritage/First+FM+Portable+Two-Way+Radio.

72 "Milestones: One-Way Police Radio Communication, 1928," *IEEE Global History Network*, n.d., http://www.ieeeghn.org/wiki/index.php/Milestones:One-Way_Police_Radio_Communication,_1928.

73 "Milestones: Two-Way Police Radio Communication, 1933," *IEEE Global History Network*, n.d., http://www.ieeeghn.org/wiki/index.php/Milestones:Two-Way_Police_Radio_Communication,_1933.

74 "1946: First Mobile Telephone Call," AT&T, n.d., http://www.corp.att.com/attlabs/reputation/timeline/46mobile.html.

75 "1946: First Mobile Telephone Call."

76 George Calhoun, *Digital Cellular Radio* (Norwood, MA: Artech House, 1988), 39.

77 "Cisco Visual Networking Index."

78 Anton Troianovski, "Video Speed Trap Lurks in New iPad," *Wall Street Journal*, last modified March 22, 2012, http://online.wsj.com/article/SB10001424052702303812904577293882009811556.html.

79 "Mobile data traffic growth doubled over one year," October 12, 2011, http://www.ericsson.com/news/111012_mobile_data_traffic_244188808_c.

80 "Mobile Network Operators Face Seven Fold Increases in Data Delivery Costs, Rising to $370 bn by 2016, Juniper Research Reports," Juniper Research, Hampshire, United Kingdom, press release, August 2, 2011, http://www.juniperresearch.com/viewpressrelease.php?pr=254.

81 Quoted in David Bollier, *Scenarios for a National Broadband Policy* (Washington, DC: Aspen Institute, 2010), http://bollier.org/sites/default/files/aspen_reports/BroadbandTEXTF_0.pdf, 9.

82 City of New York, "Frequently Asked Questions: Traffic Signs, Traffic Signals and Street Lights," http://www.nyc.gov/html/dot/html/faqs/faqs_signals.shtml, accessed September 25, 2012.

83 John Byrne, "Worldwide Cellular Infrastructure 2011–2015 Forecast: As LTE Takes Off, HSPA+ Will Remain the Technology of Choice for Many Operators," International Data Corporation, 2011, http://www.idc.com/getdoc.jsp?containerId=228061.

84 Michael Chen, "Signal Space," *Urban Omnibus*, last modified July 6, 2011, http://urbanomnibus.net/2011/07/signal-space/.

85 James E. Katz, ed. *Machines That Become Us: The Social Context of Personal Communication Technology* (New Brunswick, NJ: Transaction Publishers, 2003).

86 Quoted in John B. Kennedy, "When Woman Is Boss: An interview with Nikola Tesla by John B. Kennedy," *Collier's*, January 20, 1926.

Chapter 2. Cybernetics Redux

1 US Constitution, art. 1. sec. 2.

2 "Census of Population and Housing: 1790 Census," United States Census Bureau, U.S. Department of Commerce, http://www.census.gov/prod/www/abs/decennial/1790.html.

3 "Table 4. Population: 1790–1990," United States Census Bureau, U.S. Department of Commerce, last modified August 27, 1993, http://www.census.gov/population/censusdata/table-4.pdf.

4 "Table 4. Population: 1790–1990."

5 Campbell J. Gibson and Emily Lennon, "Historical Census Statistics on the Foreign-born Population of the United States: 1850–1990," United States Census Bureau, U.S. Department of Commerce, February 1999, http://www.census.gov/population/www/documentation/twps0029/twps0029.html.

6 "1880—History—U.S. Census Bureau," U.S. Bureau of the Census, https://www.census.gov/history/www/through_the_decades/index_of_questions/1880_1.html, accessed September 26, 2012.

7 James R. Beniger, *The Control Revolution: Technological and Economic Origins of the Information Society* (Cambridge, MA: Harvard University Press, 1986), 408–9.

8 "Census of Population and Housing: 1880 Census," United States Census Bureau, U.S. Department of Commerce, http://www.census.gov/prod/www/abs/decennial/1880.html#.

9 Emerson W. Pugh, *Building IBM: Shaping An Industry and Its Technology* (Cambridge, MA: MIT Press, 1995), 3.

10 Charles Eames and Ray Eames, *A Computer Perspective* (Cambridge, MA: Harvard University Press, 1973), quoted in Beniger, *The Control Revolution*, 411.

11 Beniger, *The Control Revolution*, vii.

12 Pugh, *Building IBM*, 4.

13 H. Hollerith, August 7, 1919, letter to J. T. Wilson, reproduced in "Historical Development of IBM Products and Patents," IBM, 1957, in Pugh, *Building IBM*, 3.

14 Pugh, *Building IBM*, 7–8.

15 Robert P. Porter, *Compendium of the Eleventh Census, Part I: Population* (Washington, DC: Government Printing Office, 1890), http://www2.census.gov/prod2/decennial/documents/1890b3_p1-01.pdf.

16 Beniger, *The Control Revolution*, 414.

17 Beniger, *The Control Revolution*, 414.

18 Pugh, *Building IBM*, 21.

19 Pugh, *Building IBM*, 4.

20 "Tabulation and Processing," United States Census Bureau, U.S. Department of Commerce, n.d., http://www.census.gov/history/www/innovations/technology/tabulation_and_processing.html.

21 Pugh, *Building IBM*, 14.

22 "A Smarter Planet The Next Leadership Agenda," November 6, 2008, video clip, Council on Foreign Relations, http://www.cfr.org/technology-and-foreign-policy/smarter-planet-next-leadership-agenda-video/p17696.

23 "IBM100-Sabre," IBM, n.d., http://www-03.ibm.com/ibm/history/ibm100/us/en/icons/sabre/.

24 "IBM100-Sabre."

25 "Colin Harrison," n.d., http://urbansystemscollaborative.org/about/leadership/colin-harrison/.

26 Colin Harrison, remarks, Ideas Economy: Intelligent Infrastructure, *The Economist* panel discussion, New York City, February 16, 2011.

27 John Tolva, telephone interview by author, November 10, 2011.

28 Tolva interview, November 10, 2011.

29 "Intelligent Cities Forum: Anne Altman," National Building Museum, last modified June 6, 2011, http://www.nbm.org/media/video/intelligent-cities/forum/intelligent-cities-forum-altman.html.

30 "IBM Deep Thunder: Frequently Asked Questions," http://www.research.ibm.com/weather/FAQs.html, accessed September 28, 2012.

31 Guru Banavar, lecture, "X-Cities 3: Heavy Weather—Design and Governance in Rio de Janeiro and Beyond," Columbia University Studio-X, New York, April 10, 2012, http://www.youtube.com/watch?v=xNsSNoL_EQM.

32 Banavar, lecture, April 10, 2012.

33 Richard J. Norton, "Feral Cities," *Naval War College Review* 56, no. 4 (2003), 105.

34 Banavar, lecture, April 10, 2012.

35 Eduardo Paes, "The 4 Commandments of Cities," TED 2012, Long Beach, California, February 29, 2012, http://www.ted.com/conversations/9659/eduardo_paes_four_commandment.html.

36 Colin Harrison, interview by author, May 9, 2011.

37 David Gelernter, *Mirror Worlds: or the Day Software Puts the Universe in a Shoebox . . . How It Will Happen and What It Will Mean* (New York: Oxford University Press, 1993), 1.

38 Gelernter, *Mirror Worlds*, 52.

39 Gelernter, *Mirror Worlds*, 5.

40 Gelernter, *Mirror Worlds*, 218.

41 Gelernter, *Mirror Worlds*, 217–18.

42 Harrison, interview, May 9, 2011.

43 Thomas Campanella, *Cities From the Sky: An Aerial Portrait of America*. (New York: Princeton Architectural Press, 2001).

44 Gelernter, *Mirror Worlds*, 222.

45 Isaac Asimov, *Foundation* (New York: Bantam Books, 2004), 17.

46 Asimov, *Foundation*, 14.

47 Paul Krugman, "Economic Science Fiction," The Conscience of a Liberal, blog, *New York Times*, last modified May 4, 2008, http://krugman.blogs.nytimes.com/2008/05/04/economic-science-fiction/.

48 Asimov, *Foundation*, 17.

49 Vannevar Bush, "As We May Think," *The Atlantic*, last modified July 1945, http://www.theatlantic.com/magazine/archive/1945/07/as-we-may-think/3881/2/.

50 Michael J. Radzicki and Robert A. Taylor. "Origin of System Dynamics: Jay W. Forrester and the History of System Dynamics" (2008), in *U.S. Department of Energy's Introduction to System Dynamics*, accessed October 23, 2008, http://www.systemdynamics.org/DL-IntroSysDyn/.

51 "2011 IW Manufacturing Hall of Fame," *Industry Week*, last modified December 11, 2011, http://www.industryweek.com/slideshows/HallofFame2011/Jay-Forrester-2011.asp.

52 Jay Wright Forrester, *Urban Dynamics* (Cambridge, MA: MIT Press, 1969), ix.

53 D. C. Lane, "The Power of the Bond Between Cause and Effect: Jay Wright Forrester and the Field of System Dynamics," *System Dynamics Review* 23, no. 2–3 (2007), 95–118.

54 G. K. Ingram, book review of *Urban Dynamics*, *Journal of the American Institute of Planners* 36, no. 3 (1970): 206–8.

55 Lincoln Quillian, "Public Housing and the Spatial Concentration of Poverty: New National Estimates," Meetings of the Population Association of America, 2005, http://paa2005.princeton.edu/download.aspx?submissionId=51567.

56 Jennifer Light, *From Warfare to Welfare: Defense Intellectuals and Urban Problems in Cold War America* (Baltimore: Johns Hopkins University Press, 2003), 47.

57 Light, *From Warfare to Welfare*, 46.

58 Douglass B. Lee Jr., "Requiem for Large-Scale Models," *Journal of the American Institute of Planners* 39, no. 3 (1973): 167.

59 Light, *From Warfare to Welfare*, 60.

60 Lee, "Requiem for Large-Scale Models," 168.

61 Joe Flood, *The Fires* (Riverhead Books: New York, 2010), 216.
62 Flood, *The Fires*, 216–17.
63 Flood, *The Fires*, 225.
64 Flood, *The Fires*, 230.
65 Flood, *The Fires*, 229.
66 Flood, *The Fires*, 18.
67 Light, *From Warfare to Welfare*, 61.
68 Lee, "Requiem for Large-Scale Models," 174.
69 Louis E. Alfeld, "Urban dynamics—the first fifty years," *System Dynamics Review* 11, no. 3 (1995): 199–217.
70 Douglass B. Lee, "Retrospective on large scale urban models," *Journal of the American Planning Association* 60, no. 1 (1994): 35–40.
71 Nicholas de Monchaux, *Spacesuit: Fashioning Apollo* (Cambridge, MA: MIT Press, 2011), 305.
72 L. Beumer, A. van Gameren, B. van der Hee, and J. Paelinck, "A Study of the Formal Structure of J. W. Forrester's Urban Dynamics Model," *Urban Studies* 15 (1978): 167.
73 Light, *From Warfare to Welfare*, 58.
74 Joe Zehnder, telephone interview by author, August 29, 2012.
75 "IBM Smarter City: Portland, Oregon," *YouTube* video, August 12, 2011, http://www.youtube.com/watch?v=uBYsSFbBeR4.
76 Justin Cook, telephone interview by author, September 11, 2012.
77 Zehnder, interview, August 29, 2012.
78 Cook, interview, September 11, 2012.
79 Cook, interview, September 11, 2012.
80 Zehnder, interview, August 29, 2012.
81 Zehnder, interview, August 29, 2012.
82 Michael Batty, "Building a science of cities," *Cities*, 2011, doi:10.1016/j.cities.2011.11.008, 1.
83 *All Watched Over By Machines of Loving Grace*, directed by Adam Curtis (2011; BBC).
84 Jay Forrester, "System Dynamics and the Lessons of 35 Years," in Kenyon B. De Greene, *A Systems-Based Approach to Policymaking* (Boston: Kluwer Academic Publishers, 1993), 202.
85 Alfeld, "Urban dynamics—the first fifty years."
86 B. Raney et al., "An agent-based microsimulation model of Swiss travel: First results," *Networks and Spatial Economics* 3, no. 1 (2003): 23–42.
87 Michael Batty, telephone interview by author, August 19, 2010.

88 "Heisenberg-Quantum Mechanics, 1925–1927: The Uncertainty Principle," American Institute of Physics, n.d., accessed February 26, 2013, http://www.aip.org/history/heisenberg/p08.htm.

89 Asimov, *Foundation*, 14.

90 Lee, "Requiem for Large-Scale Models," 167.

91 Harrison, interview, May 9, 2011.

92 Gelernter, *Mirror Worlds*, 217, Gelernter's italics.

93 "SimCity and Advanced GeoAnalytics," *SpatialMarkets* blog, March 16, 2012, http://www.spatialmarkets.com/2012/3/16/simcity-and-advanced-geoanalytics.html.

94 Lee, "Requiem for Large-Scale Models," 169.

95 Banavar, lecture, April 10, 2012.

96 Gelernter, *Mirror Worlds*, 222.

Chapter 3. Cities of Tomorrow

1 Ebenezer Howard, *Garden Cities of To-morrow* (London: Swan Sonnenschein & Co., Ltd., 1902), 18–26.

2 Robert H. Kargon and Arthur P. Molella, *Invented Edens: Techno-Cities of the 20th Century* (Cambridge, MA: MIT Press), 24.

3 Kargon and Molella, *Invented Edens*, 18.

4 Volker Welter, *Biopolis: Patrick Geddes and the City of Life* (Cambridge, MA: MIT Press, 2003), 11.

5 Patrick Geddes, *Civics as Applied Sociology* (Middlesex, UK: The Echo Library, 2008), 5.

6 Jane Jacobs, *The Death and Life of Great American Cities* (New York: Random House, 1961), 19.

7 Robert Fishman, "The Death and Life of Regional Planning," in *Reflections on Regionalism*, edited by B. Katz (Washington, DC: Brookings Institution, 2000), 115. Fishman's original source material is Jacobs, *Death and Life of Great American Cities*, chap. 7.

8 Thomas J. Campanella, "Jane Jacobs and the Death and Life of American Planning," *Places: Forum of Design for the Public Realm*, April 25, 2011, http://places.designobserver.com/feature/jane-jacobs-and-the-death-and-life-of-american-planning/25188/.

9 Campanella, "Jane Jacobs and the Death and Life of American Planning."

10 R. L. Duffus, "A Rising Tide of Traffic Rolls Over New York; What is Being Done to Relieve the Ever-Growing Street Congestion Which Threatens to

Slow Up the Vital Processes of Life in the Metropolis," *New York Times*, February 9, 1930, XX4.

11 Peter D. Norton, *Fighting Traffic: The Dawn of the Motor Age in the American City* (Cambridge, MA: MIT Press, 2008), 25.

12 Duffus, "A Rising Tide of Traffic Rolls Over New York," XX4.

13 Norton, *Fighting Traffic*, 25–27.

14 Norton, *Fighting Traffic*, 24.

15 Norton, *Fighting Traffic*, 105.

16 Norton, *Fighting Traffic*, 2.

17 Campanella, "Jane Jacobs and the Death and Life of American Planning."

18 Anthony Flint, *Wrestling with Moses: How Jane Jacobs Took on New York's Master Builder and Transformed the American City* (New York: Random House, 2009), 51.

19 Author's calculation using estimates from Caro, *The Power Broker*, 9, and US Bureau of Labor Statistics CPI Inflation Calculator, http://www.bls.gov/cpi/cpicalc.htm, accessed August 15, 2012.

20 Flint, *Wrestling with Moses*, 85–87.

21 Flint, *Wrestling with Moses*, 100.

22 Flint, *Wrestling with Moses*, 105.

23 Flint, *Wrestling with Moses*, 99.

24 Flint, *Wrestling with Moses*, 109.

25 Campanella, "Jane Jacobs and the Death and Life of American Planning."

26 Tom Wright, remarks, "Tools for Engagement" workshop, Regional Plan Association & Lincoln Institute for Land Policy, New York, March 29, 2012.

27 Patrick Geddes, quoted in Jaqueline Tyrwhitt, ed., *Patrick Geddes in India* (London: Lund Humphries: 1947), 45.

28 Helen Meller, *Patrick Geddes: Social Evolutionist and City Planner* (New York: Routledge, 1990), 76–79.

29 Patrick Geddes, quoted in Tyrwhitt, ed., *Patrick Geddes in India*, 41

30 Alasdair Geddes, quoted in Tyrwhitt, ed., *Patrick Geddes in India*, 15.

31 Lewis Mumford, quoted in in Tyrwhitt, ed., *Patrick Geddes in India*, 11.

32 Quoted in Welter, *Biopolis*, 18.

33 Nicolai Ouroussoff, "Outgrowing Jane Jacobs and Her New York, *New York Times*, April 30, 2006, http://www,nytimes.com/2006/04/30/weekinreview/30jacobs.html.

34 Campanella, "Jane Jacobs and the Death and Life of American Planning."

35 Fareed Zakaria, "Special Address: At the Intersection of Globalization and Urbanization," SmarterCities Forum, Rio de Janeiro, Brazil, November 9, 2011.

36 Tyler Cowen, *The Great Stagnation: How America Ate All The Low-Hanging*

Fruit of Modern History, Got Sick, and Will (Eventually) Feel Better (New York: Dutton, 2011), Kindle edition, location 93.

37 "Hal Varian on How the Web Challenges Managers," video interview with James Manyika, McKinsey & Co., last modified January 2009, http://www.mckinseyquarterly.com/Hal_Varian_on_how_the_Web_challenges_managers_2286.

38 Joi Ito, "The Internet, innovation and learning," last modified December 5, 2011, http://joi.ito.com/weblog/2011/12/05/the-internet-in.html.

39 Ito, "The Internet, innovation and learning."

40 Michael Hiltzik, "So, who really did invent the Internet?" *Los Angeles Times*, last modified July 23, 2012, http://www.latimes.com/business/money/la-mo-who-invented-internet-20120723,0,5052169.story.

41 Bernard Rudofsky, *Architecture Without Architects* (Albuquerque: University of New Mexico Press, 1987).

42 Gary Wolf, "Exploring the Unmaterial World," *Wired*, 2000, 306–19.

43 Gene Becker, "Prada Epicenter Revisited," *Fred's House*, blog, last modified April 4, 2004, http://www.fredshouse.net/archive/000159.html.

44 Adam Greenfield, *Everyware: The Dawning Age of Ubiquitous Computing* (Berkeley, CA: New Riders, 2006), 179.

45 M. Weiser, "Ubiquitous Computing," last modified March 17, 1996, accessed August 18, 2012, http://web.archive.org/web/20070202035810/http://www.ubiq.com/hypertext/weiser/UbiHome.html.

46 Meller, *Patrick Geddes: Social Evolutionist and City Planner*, 143–44.

47 Lewis Mumford, "Mumford on Geddes," *The Architectural Review* 108, no. 644 (1950): 86-7.

Chapter 4. The Open-Source Metropolis

1 Red Burns, "Cultural Identity and Integration in the New Media World," paper presented at University of Industrial Arts, Helsinki, Finland, November 19–21, 1991.

2 "United States: Cable Television," Museum of Broadcast Communications, n.d., http://www.museum.tv/eotvsection.php?entrycode=unitedstatesc.

3 "History of Cable Television," National Cable & Telecommunications Association, n.d., http://www.ncta.com/About/About/HistoryofCableTelevision.aspx.

4 National Cable & Telecommunications Association, n.d., retrieved from Internet Archive, http://web.archive.org/web/20120103181806/http://www.ncta.com/About/About/HistoryofCableTelevision.aspx?source=Resources.

5 "History of Cable Television."

6 Jason Huff, "Technology is Not Enough: The Story of NYU's Interactive Telecommunications Program," Rhizome, December 15, 2011, http://rhizome.org/editorial/2011/dec/15/technology-not-enough-story-nyus-interactive-telec/.

7 Red Burns, original manuscript, "Beyond Statistics," Alternate Media Center, School of the Arts, New York University, n.d., 7. Also published in Martin C. J. Elton et. al., eds., *Evaluating New Telecommunications Services* (New York: Plenum Publishing, 1978).

8 Burns, "Cultural Identity and Integration in the New Media World," 6–7.

9 Red Burns, interview by author, New York, October 24, 2011.

10 Martin Elton, martin.elton@nyu.edu, "Through the Looking Glass: The Rhizome article on ITP," private e-mail reposted by Gilad Rosner, itp-alumni@lists.nyu.edu, December 21, 2011.

11 Red Burns, "Technology is not enough," paper presented at the American Council on Education, Washington, DC, October 16, 1981.

12 Burns, interview, October 24, 2011.

13 William Gibson, "Rocket Radio," *Rolling Stone*, June 15, 1989.

14 Burns, "Cultural Identity and Integration in the New Media World," 7.

15 Dennis Crowley, interview by author, May 13, 2011.

16 Dodgeball.com, November 9, 2000, http://web.archive.org/web/20001109 2025/http://www.dodgeball.com/city/.

17 Crowley, interview, May 13, 2011.

18 Crowley, interview, May 13, 2011.

19 Crowley, interview, May 13, 2011.

20 Five years later, when a half-decade's worth of archived check-ins were migrated to the Foursquare database, Crowley e-mailed with the discovery that on November 17, 2003, during a test, I had tapped out a terse message, hit send, and became the first person (other the Crowley and Rainert) to check in on the third version of Dodgeball.

21 Crowley, interview, May 13, 2011.

22 Laura Barnett, "If It Wasn't For Hedy Lamarr, We Wouldn't Have Wi-Fi," *The Guardian*, last modified December 4, 2011, http://www.guardian.co.uk/theguardian/shortcuts/2011/dec/04/hedy-lamarr-wifi.

23 "A Brief History of Wi-Fi," *The Economist*, June 10, 2004, http://www.economist.com/node/2724397.

24 Alvin F. Harlow, *Old Wires and New Waves: The History of the Telegraph, Telephone and Wireless* (New York: D. Appleton-Century, 1936), 456; "About IIT:

Hall of Fame: Lee DeForest," last modified October 2, 2012, http://www.iit.edu/about/history/hall_of_fame/lee_de_forest.shtml; SCANFAX Year in Review, "Lee de Forest: Father of Radio Broadcasting and Receiving," IEEE-Chicago Section: Chicago, Illinois, 2008, 13, http://www.ieeechicago.org/LinkClick.aspx?fileticket=X8F8-rFhkPY%3D&tabid=421.

25 Rob Flickenger, "Antenna on the Cheap (er, Chip)," O'Reilly Wireless DevCenter, blog, last modified July 5, 2001, http://www.oreillynet.com/cs/weblog/view/wlg/448.

26 Untitled broadcast, CNN Moneyline, June 8, 2001, transcript available at http://www.cnn.com/TRANSCRIPTS/0106/08/mlld.00.html.

27 Thor Olavsrud, "Intel, IBM Team With AT&T To Push Nationwide Wi-Fi," last modified December 5, 2002, http://www.internetnews.com/wireless/article.php/1553001/Intel+IBM+Team+With+ATT+To+Push+Nationwide+WiFi.htm.

28 Clark Boyd, "Estonia's 'Johnny Appleseed' of Free Wi-Fi," *Discovery News*, last modified July 11, 2010, http://news.discovery.com/tech/estonias-johnny-appleseed-of-free-wi-fi.html; Si Hawkins, "Tallinn: City of the Future," EasyJet Traveller, February 11, 2011, http://traveller.easyjet.com/features/2011/02/tallinn-city-of-the-future.

29 Hamish McKenzie, "Dennis Crowley: Google Acquisition of Dodgeball A Failure," *PandoDaily*, blog, last modified October 11, 2012, http://pandodaily.com/2012/10/11/foursquares-dennis-crowley-google-acquisition-of-dodgeball-a-failure/.

30 "Botanicalls: The Plants Have Your Number," Botanicalls website, accessed February 10, 2012, http://www.botanicalls.com/classic/.

31 "Botanicalls: Plants Have Your Number," July 7, 2008, http://www.youtube.com/watch?v=mqzwru0sQY4.

32 "Daniel Rozin Wooden Mirror," n.d., http://www.smoothware.com/danny/woodenmirror.html.

33 Tom Igoe, interview by author, October 6, 2011.

34 Phillip Torrone, "Why the Arduino Won and Why It's Here to Stay," *Make*, blog, last modified February 10, 2011, http://blog.makezine.com/2011/02/10/why-the-arduino-won-and-why-its-here-to-stay/.

35 Clive Thompson, "Build It. Share It. Profit. Can Open Source Hardware Work?," *Wired*, October 20, 2008, http://www.wired.com/techbiz/startups/magazine/16-11/ff_openmanufacturing.

36 Torrone, "Why the Arduino Won and Why It's Here to Stay."

37 Igoe, interview, October 6, 2011.

38 Torrone, "Why the Arduino Won and Why It's Here to Stay."
39 Igoe, interview, October 6, 2011.
40 Riverkeeper, "Combined Sewage Outflows (CSOs)," accessed September 24, 2012, http://www.riverkeeper.org/campaigns/stop-polluters/cso/.
41 Victoria Bekiempis, "Sewage Secrets: Leif Percifield Does Not Like It Raw," *Village Voice*, blog, last modified January 23, 2012, http://blogs.villagevoice.com/runninscared/2012/01/sewage_secrets.php.
42 Torrone, "Why the Arduino Won and Why It's Here to Stay."
43 Igoe, interview, October 6, 2011.
44 Igoe, interview, October 6, 2011.

Chapter 5. Tinkering Toward Utopia

1 Christopher Alexander, "A City is Not A Tree," *Architectural Forum* 122, no. 1 (1965): 58–62.
2 Alexander, "A City is Not A Tree."
3 Doug Lea, "Christopher Alexander: An Introduction for Object-Oriented Designers," *Software Engineering Notes* 19, no. 1 (1994): 39–46, http://www.ics.uci.edu/~andre/informatics223s2011/lea.pdf. See also Subrata Dasgupta, *Design Theory and Computer Science* (Cambridge: Cambridge University Press, 1991).
4 Nicholas Carson, "15 Google Interview Questions That Will Make You Feel Stupid," *BusinessInsider*, last modified November 4, 2009, http://www.businessinsider.com/15-google-interview-questions-that-will-make-you-feel-stupid-2009-11.
5 Pascal-Emmanuel Gobry, "Foursquare Gets 3 Million Check-Ins Per Day, Signed Up 500,000 Merchants," *BusinessInsider*, last modified August 2, 2011, http://articles.businessinsider.com/2011-08-02/tech/30097137_1_foursquare-users-merchants-ins.
6 Kori Schulman, "Take A Tip From the White House on Foursquare," *The White House*, blog, last modified August 15, 2011, http://www.whitehouse.gov/blog/2011/08/15/take-tip-white-house-foursquare.
7 Dennis Crowley, interview by author, May 13, 2011.
8 Crowley, interview, May 13, 2011.
9 Liz Gannes, "Foursquare's Version of the Talent Acquisition: Summer Interns," *All Things D*, blog, last modified July 1, 2011, http://allthingsd.com/20110701/foursquares-version-of-the-talent-acquisition-summer-interns/.
10 Ingrid Lunden, "Foursquare's Inflection Point: People Using The App, But Not Checking In," *Tech Crunch*, last modified March 2, 2012, http://tech

crunch.com/2012/03/02/foursquares-inflection-point-people-using-the-app-but-not-checking-in/.

11 Matthew Flamm, "Foursquare Doesn't Quite Check Out," *Crain's New York Business*, January 20, 2013, http://www.crainsnewyork.com/article/20130120/TECHNOLOGY/301209972.

12 H. Edward Roberts and William Yates, "ALTAIR 8800: The most powerful minicomputer project ever presented—can be built for under $400," *Popular Electronics*, January 1975, 33.

13 Steve Ditlea, ed., *Digital Deli: The Comprehensive, User-lovable Menu of Computer Lore, Culture, Lifestyles and Fancy* (New York: Workman, 1984), 74–75.

14 People's Computer Network, "Newsletter #1," October 1972, http://www.digibarn.com/collections/newsletters/peoples-computer/peoples-1972-oct/index.html.

15 Ian Keldoulis, "Where Good Wi-Fi Makes Good Neighbors," *New York Times*, last modified October 21, 2004, http://www.nytimes.com/2004/10/21/technology/circuits/21spot.html?_r=1&ex=1256097600&en=4ed99f1b6f6cb878&ei=5090&partner=rssuserland.

16 Fred Wilson, "Meetups," *AVC*, blog, last modified April 17, 2008, http://www.avc.com/a_vc/2008/04/meetups.html.

17 Whitney McNamara, "Anatomy of A Twitter Bot," *Seamonkeyradio*, blog, last modified April 22, 2008, http://smr.absono.us/2008/04/anatomy-of-a-twitter-bot/.

18 DIYcity, "DIYcity: How do you want to reinvent your city?," last modified July 25, 2010, http://www.icyte.com/system/snapshots/fs1/0/5/6/2/05625d480d276043326229910d11701abae39965/index.html.

19 "About" DIYcity, n.d., http://diycity.org/about.

20 John Geraci, interview by author, November 1, 2011.

21 Geraci, interview, Novermber 1, 2011.

22 Scott Heiferman, "9/11 & us," *Meetup HQ*, blog, last modified September 9, 2011, http://meetupblog.meetup.com/post/21449652035/9-11-us.

23 Edward Glaeser, *Triumph of the City* (New York: Penguin Press, 2011), 128.

24 Geoffrey West, lecture at Urban Systems Symposium, New York University, New York, May 12, 2011.

25 Kevin Lynch, *The Image of the City* (Cambridge, MA: MIT Press, 1960), 126.

26 Mitchell L. Moss, "Telecommunications, World Cities, and Urban Policy," *Urban Studies*, December 1987.

27 F. Calabrese and F. and C. Ratti, "Real Time Rome," *Networks and Communications Studies* 20, no. 3–4 (2006): 247–58.

28 J. Borge-Holthoefer et al., "Structural and Dynamical Patterns on Online

Social Networks: The Spanish May 15th Movement as a Case Study," *PLoS ONE*, (2011); doi:10.1371/journal.pone.0023883.

29 April Kilcrease, "A Conversation with Zipcar's CEO Scott Griffith," *GigaOM*, last modified December 5, 2011, http://gigaom.com/cleantech/a-conversation-with-zipcars-ceo-scott-griffith/.

30 Ron Lieber, "Share Your Car, Risk Your Insurance," *New York Times*, last modified March 16, 2012, http://www.nytimes.com/2012/03/17/your-money/auto-insurance/enthusiastic-about-car-sharing-your-insurer-isnt.html?pagewanted=all.

31 "Our Carbon Footprint," *Corporate Responsibility Report*, InterContinental Hotel Groups, 2011, http://www.ihgplc.com/index.asp?pageid=747.

32 *Building Design and Construction: Forging Resource Efficiency and Sustainable Development*, United Nations Environment Programme, Nairobi, Kenya, June 2012, https://www.usgbc.org/ShowFile.aspx?DocumentID=19073.

33 Frank Duffy, *Work and the City* (London: Black Dog Publishing, 2008).

34 Red Burns, interview by author, October 24, 2011.

35 Geraci, interview, November 1, 2011.

36 Burns, interview, October 24, 2011.

Chapter 6. Have Nots

1 For an excellent discussion of the role of Twitter in the Moldovan revolution in 2009, see Evgeny Morozov, "Moldova's Twitter revolution is NOT a myth," *Foreign Policy NET.EFFECT*, blog, last modified April 10, 2009, http://neteffect.foreignpolicy.com/posts/2009/04/10/moldovas_twitter_revolution_is_not_a_myth.

2 *Moldova Economic Sector Analysis: Final Report*, U.S. Agency for International Development: Washington, DC, March 2010, http://pdf.usaid.gov/pdf_docs/PNADU233.pdf.

3 AnnaLee Saxenian, *The New Argonauts: Regional Advantage in a Global Economy* (Cambridge, MA: Harvard University Press, 2007).

4 Plato, *The Republic*, translated by Benjamin Jowett, The Internet Classics Archive, http://classics.mit.edu/Plato/republic.html, accessed December 5, 2012.

5 "The Challenge," UN Habitat, n.d., http://www.unhabitat.org/content.asp?typeid=19&catid=10&cid=928.

6 Richard Heeks, "ICT4D 2.0: The Next Phase of Applying ICT for International Development," *IEEE Computer* 41, no. 6 (2008): 27.

7 Heeks, "ICT4D 2.0," 27.

8 J. M. Figueres, A. Cruz, J. Barrios, and A. Pentland, "A Practical Plan: The Little Intelligent Communities Project," n.d., http://www.media.mit.edu/unwired/theproject.html.

9 Paul Brand and Anke Schwittay, "The Missing Piece: Human-Driven Design and Research in ICT and Development," International Conference on Information Communication Technology and Development, 2006, http://www.qatar.cmu.edu/iliano/courses/07F-CMU-CS502/papers/Brand-and-Schwittay.pdf. See also M. Granqvist, "Looking critically at ICT4Dev: The Case of Lincos," *The Journal of Community Informatics* 2, no.1 (2005).

10 Alice Rawthorn, "A Few Stumbles on the Road to Connectivity" *New York Times*, last modified December 19, 2011, http://www.nytimes.com/2011/12/19/arts/design/a-few-stumbles-on-the-road-to-connectivity.html.

11 "The World in 2011: ICT Facts and Figures," International Telecommunications Union, Geneva, 2011, http://www.itu.int/ITU-D/ict/facts/2011/material/ICTFactsFigures2011.pdf.

12 "Spotlight on Africa—Mobile Statistics & Facts 2012," video clip, Youtube, last modified July 9, 2012, https://www.youtube.com/watch?v=0bXjgx4J0C4&feature=player_embedded.

13 Killian Fox, "Africa's mobile economic revolution," *The Observer*, July 23, 2011, http://www.guardian.co.uk/technology/2011/jul/24/mobile-phones-africa-microfinance-farming.

14 "Celebrate the IDEOS vs. Samsung $100 Smartphone Price War in Kenya," Inveneo ICTworks, last modified July 20, 2012, http://www.ictworks.org/news/2012/07/20/celebrate-ideos-vs-samsung-100-smartphone-price-war-kenya.

15 Jon Evans, "In Five Years, Most Africans Will Have Smartphones," TechCrunch, blog, last modified June 9, 2012, http://techcrunch.com/2012/06/09/feature-phones-are-not-the-future/.

16 Christine Zhen-Wei Qiang, "Mobile Telephony: A Transformational Tool for Growth and Development," *Private Sector & Development*, Proparco, November 2009, http://www.ffem.fr/jahia/webdav/site/proparco/shared/PORTAILS/Secteur_prive_developpement/PDF/SPD4_PDF/Christine-Zhen-Wei-Qiang-World-Bank-Mobile-Telephony-A-Transformational-Tool-for-Growth-and-Development.pdf.

17 Nancy Odendaal, lecture, Forum on Future Cities, MIT SENSEable City Lab and the Rockefeller Foundation, Cambridge, MA, April 12, 2011, http://techtv.mit.edu/collections/senseable/videos/12306-changing-government.

18 For the cost of building fiber-optic networks: Erin Bohlin, Simon Forge, and

Colin Blackman, "Telecom Infrastructure to 2030," in *Infrastructure to 2030: Telecom, Land Transport, Water and Electricity*, Organisation for Economic Co-operation and Development (Paris: OECD Publishing 2006), 90; for the cost of wireless broadband networks: Pulak Chowdhury, Suman Sarkar, and Abu Ahmed (Sayeem) Reaz, "Comparative Cost Study of Broadband Access Technologies," University of California, Davis, Department of Computer Science, n.d, http://networks.cs.ucdavis.edu/~pulak/papers/broadband_cost_study_ANTS.pdf.

19 "Broader 4G wireless access will accelerate economic development and improve quality of life in rural and developing regions of the world, say IEEE wireless experts," *Express Computer Online*, n.d., http://www.expresscomputeronline.com/20110615/news21.shtml.

20 "Startups in Bangalore: Babajob," *Podtech*, blog, last modified September 5, 2007, http://www.podtech.net/home/4043/startups-in-bangalore-babajob.

21 Ayesha Khanna, "Is your city smart enough?" *Indian Express*, last modified January 3, 2012, http://www.indianexpress.com/news/is-your-city-smart-enough/894919/.

22 "UN award for SA's Dr Math mobile tool," *SouthAfrica.info*, blog, last modified June 9, 2011, http://www.southafrica.info/business/trends/innovations/drmath-090611.htm#.UHA-00IQTzI.

23 Katrina Manson, "Kenya to India: exporting the mobile money model," *Financial Times*, blog, last modified November 11, 2011, http://blogs.ft.com/beyond-brics/2011/11/11/kenya-to-india-exporting-the-mobile-money-model/.

24 "Ericsson and Orange bring sustainable and affordable connectivity to rural Africa," Telefonaktiebolaget LM Ericsson, Stockholm, last modified February 18, 2009, http://www.ericsson.com/news/1291529.

25 Andrew Nusca, "Vodafone Debuts $32 Solar-Powered Mobile Phone for Rural India," *Smart Planet*, blog, last modified July 27, 2010, http://www.smartplanet.com/blog/smart-takes/vodafone-debuts-32-solar-powered-mobile-phone-for-rural-india/9367.

26 A. Wesolowski and N. Eagle, "Parameterizing the Dynamics of Slums," AAAI Spring Symposium 2010 on Artificial Intelligence for Development (AI-D), 2010, http://ai-d.org/pdfs/Wesolowski.pdf.

27 Mirjam E. de Bruijn, "Mobile Telephony and Socio-Economic Dynamics in Africa," in *Global Infrastructure: Ongoing realities and emerging challenges*, edited by Gregory K. Ingram and Karin L. Brandt (Cambridge, MA: Lincoln Institute for Land Policy, forthcoming 2013).

28 Heeks, "ICT4D 2.0," 28.

29 Eric Schmidt, "A Week of Africa," January 22, 2013, https://plus.google.com/+EricSchmidt/posts/VRFReMyLwfS.

30 Quotes in this section are from Robert Kirkpatrick and Ban Ki-moon: remarks, United Nations General Assembly, New York, November 8, 2011.

31 "Agile Global Development: Harnessing the Power of Real-Time Information," *Global Pulse*, Fall 2011, http://uscpublicdiplomacy.org/media/GlobalPulseFall2011.pdf.

32 Gregory T. Huang, "Jana, Formerly Txteagle, Unveils Strategy for 'Giving 2 Billion People a Raise'—A Talk with CEO Nathan Eagle," *Xconomy*, blog, last modified October 11, 2011, http://www.xconomy.com/boston/2011/10/11/jana-formerly-txteagle-unveils-strategy-for-giving-2-billion-people-a-raise-a-talk-with-ceo-nathan-eagle/.

33 "Global Snapshot of Well-Being—Mobile Survey," *UN Global Pulse project* website, n.d., http://www.unglobalpulse.org/projects/global-snapshot-well being-mobile-survey.

34 Megan Lane, "As Asbo in 14th Century Britain," *BBC News Magazine*, April 5, 2011, http://www.bbc.co.uk/news/magazine-12847529.

35 Martin Daunton, "London's 'Great Stink' and Victorian Urban Planning," *BBC History*, November 4, 2004, http://www.bbc.co.uk/history/trail/victorian_britain/social_conditions/victorian_urban_planning_04.shtml.

36 C. Creighton, *A History of Epidemics in Britain* (Oxford: Oxford University Press, 1894), 858.

37 While the official 2009 Kenya census tallied Kibera's population at 170,070, two other estimates put the figure closer to 250,000. One extrapolated from a door-to-door survey in one of the slum's districts, the other used satellite imagery to count structures. See Mikel Maron, "Kibera's Census: Population, Politics, Precision," September 5, 2010, http://www.mapkibera.org/blog/2010/09/05/kiberas-census-population-politics-precision/.

38 Pratima Joshi, Srinanda Sen, and Jane Hobson, "Experiences with surveying and mapping Pune and Sangli slums on a geographical information system (GIS)," *Environment & Urbanization* 14, no. 2 (2002): 225, http://www.ucl.ac.uk/dpu-projects/drivers_urb_change/urb_governance/pdf_comm_act/IIED_Joshi_Hobson_Pune_GIS.pdf.

39 Mikel Maron, telephone interview by author, March 27, 2012.

40 Charles Arthur, "Ordnance Survey launches free downloadable maps," *The Guardian*, March 31, 2010, http://www.guardian.co.uk/technology/2010/apr/01/ordnance-survey-maps-download-free.

41 Maron, interview, March 27, 2012.

42 Erica Hagen, "The story of Map Kibera," IKM Emergent program wiki, http://wiki.ikmemergent.net/index.php/Workspaces:The_changing_environment_of_infomediaries/Map_Kibera, accessed March 17, 2012.

43 Heeks, "ICT4D 2.0," 30.

44 P. Joshi, S. Sen, and J. Hobson, "Experiences with surveying and mapping Pune and Sangli slums on a geographical information system (GIS)," 225.

45 *State of the World's Cities Report 2008/9: Harmonious Cities*, (Nairobi, Kenya: UN-HABITAT, 2008).

46 Victor Mulas, "Why broadband does not always have an impact on economic growth?," World Bank Information and Communications for Development, blog, last modified November 21, 2011, http://blogs.worldbank.org/ic4d/why-broadband-does-not-always-have-an-impact-on-economic-growth.

47 Steven Johnson, "What a Hundred Million Calls to 311 Reveal About New York," *Wired*, last modified November 1, 2010, http://www.wired.com/magazine/2010/11/ff_311_new_york/all/1.

48 Sarah Williams and Nick Klein, "311 Complaint Spatial Analysis Assessment," report to New York City Department of Sanitation, November 2007, http://www.s-e-w.net/DSNY/DSNYfinalreport_ver2.pdf.

49 See for instance, "Multidimensional Poverty Index," Oxford Poverty and Human Development Initiative, Department of International Development, University of Oxford, 2011, http://www.ophi.org.uk/policy/multidimensional-poverty-index/.

50 Richard Heeks, "The ICT4D 2.0 Manifesto: Where Next for ICTs and International Development?," Working Paper No. 42, Institute for Development Policy and Management, University of Manchester, http://www.oecd.org/ict/4d/43602651.pdf.

51 Erik Hersman, "From Kenya to Madagascar: The African tech-hub boom," *BBC News Business*, blog, last modified July 19, 2012, http://www.bbc.co.uk/news/business-18878585.

Chapter 7. Reinventing City Hall

1 Bob Tedeschi, "Big Wi-Fi Project for Philadelphia," *New York Times*, September 27, 2004, http://www.nytimes.com/2004/09/27/technology/27ecom.html.

2 Brian James Kirk and Christopher Wink, "The Wireless Philadelphia Problem," Technical Philly, n.d., http://technicallyphilly.com/dp/wptimeline.swf.

3 Greg Goldman quotes from lecture, Wireless City: Can All New Yorkers Get Connected?, Municipal Art Society of New York, last modified March 11, 2011, http://mas.org/wireless-city-can-all-new-yorkers-get-connected-panel-video/.

4 Tom McGrath, "The Next Great American City: It might just be us. Philadelphia. What, exactly, is going on?" *Philadelphia Magazine*, December 2005, http://www.phillymag.com/articles/features-the-next-great-american-city/.

5 "Municipal Wi-Fi Networks Run Into Financial, Technical Trouble," *Associated Press*, Fox News, last modified May 23 2007, http://www.foxnews.com/story/0,2933,274728,00.html.

6 Ian Urbina, "Hopes for Wireless Cities Fade as Internet Providers Pull Out," *New York Times*, last modified March 22, 2008 http://www.nytimes.com/2008/03/22/us/22wireless.html?pagewanted=all.

7 Siddhartha Mahanta, "Why Are Telecom Companies Blocking Rural America From Getting High-Speed Internet?" *The New Republic*, last modified April 17, 2012, http://www.tnr.com/article/politics/102699/rural-broadband-internet-wifi-access.

8 Jon Leibowitz, "Municipal Broadband: Should Cities Have a Voice?," National Association of Telecommunications Officers and Advisors (NATOA) 25th Annual Conference, September 22, 2005, http://www.ftc.gov/speeches/leibowitz/050922municipalbroadband.pdf.

9 Jefferson Dodge, "Terrifying Telecom Tale: Corporations Bankrolling Fight Against Local Network Measure—Again," *Boulder Weekly*, October 20, 2011, http://www.boulderweekly.com/article-6722-terrifying-telecom-tale.html.

10 Institute for Local Self-Reliance, "Community Broadband Bits 10—Vince Jordan from Longmont, Colorado," last modified August 28, 2012, http://www.muninetworks.org/content/community-broadband-bits-10-vince-jordan-longmont-colorado.

11 "City to buy what's left of Wireless Philadelphia for $2 million," Philly.com, last modified December 16, 2009, http://www.philly.com/philly/blogs/heardinthehall/79437182.html.

12 Juan Gonzalez, "City couldn't sell back fancy wireless," *New York Daily News*, last modified February 15, 2012, http://articles.nydailynews.com/2012-02-15/news/31064869_1_public-safety-agencies-system-public-safety.

13 Anick Jesdanun, "Cities struggle with wireless internet," *USA Today*, last modified May 22, 2007, http://www.usatoday.com/tech/products/2007-05-21-297466529_x.htm.

14 "47 Applications in 30 Days for $50K," Government Technology, last modi-

fied November 13, 2008, http://www.govtech.com/e-government/47-Applications-in-30-Days-for.html.

15 "Apps for Democracy Yields 4,000% ROI in 30 Days for DC.Gov," iStrategy Labs, last modified November 25, 2008, http://istrategylabs.com/2008/11/apps-for-democracy-yeilds-4000-roi-in-30-days-for-dcgov/.

16 Trees Near You, n.d., http://www.treesnearyou.com/.

17 John Geraci, interview by author, November 3, 2010.

18 Hana Schank, "New York's Digital Deficiency," *Fast Company*, last modified December 14, 2011, http://www.fastcompany.com/1800674/new-york-citys-digital-deficiency.

19 Steven Towns, "Government 'Apps' Move from Cool to Useful," *Governing*, last modified June 7, 2010, http://www.governing.com/columns/tech-talk/Government-Apps-Move-from.html.

20 Gautham Nagesh, "New D.C. CTO Scraps 'Apps for Democracy,'" *The Hill*, blog, last modified June 7, 2010, http://thehill.com/blogs/hillicon-valley/technology/101779-new-dc-cto-scraps-apps-for-democracy.

21 Matthew Roth, "How Google and Portland's TriMet set the Standard for Open Transit Data," San Francisco Streets, blog, last modified January 5, 2010, http://sf.streetsblog.org/2010/01/05/how-google-and-portlands-trimet-set-the-standard-for-open-transit-data/.

22 Francisca Rojas, telephone interview by author, November 15, 2011.

23 Hills are Evil!, n.d., http://www.hillsareevil.com/.

24 Tom Olmstead, "How New York is Going Digital in 2011," *Mashable*, last modified June 22, 2011, http://mashable.com/2011/06/22/new-york-digital-rachel-sterne/.

25 Dirk Johnson, "In Privatizing City Services, It's Now 'Indy-a-First-Place,'" *New York Times*, March 2, 1995, http://www.nytimes.com/1995/03/02/us/in-privatizing-city-services-it-s-now-indy-a-first-place.html.

26 Sewell Chan, "Remembering a Snowstorm that Paralyzed a City," *New York Times*, blog, last modified February 10, 2009, http://cityroom.blogs.nytimes.com/2009/02/10/remembering-a-snowstorm-that-paralyzed-the-city/.

27 At the time, Goldsmith's stated reason for resigning was "to return to academic work and to pursue opportunities in the financial sector," the *New York Times* reported, but it was later disclosed that he had been arrested in Washington, DC, for domestic violence on July 30, 2011. However, the fallout from the blizzard fiasco had long since rendered him ineffective, and his departure coincided with the departure of several other high-profile outsiders brought in to run the third Bloomberg administration.

28 John Byrne, "Daley deflects LSD criticism, says officials did 'very good job,'" *Chicago Tribune*, last modified February 3, 2011, http://www.chicagotribune.com/news/local/breaking/chibrknews-cars-cleared-from-lake-shore-drive-which-remains-closed-20110203,0,7546399.story.

29 David Ariosto, "New York to tag snow plows with GPS after clean-up controversy," CNN, last modified January 6, 2011, http://articles.cnn.com/2011-01-06/us/new.york.bloomberg.snow_1_snow-plows-officials-brace-clean-up-efforts?_s=PM:US; "Chicago tracks fleet vehicles on Web-based maps," American City and County, last modified June 1, 2003, http://americancityandcounty.com/mag/government_chicago_tracks_fleet.

30 Gerald F. Seib, "In Crisis, Opportunity for Obama," *Wall Street Journal*, November 21, 2008, http://online.wsj.com/article/SB122721278056345271.html.

31 John Tolva, telephone interview by author, November 10, 2011.

32 Tolva, interview, November 10, 2011.

33 Tolva, interview, November 10, 2011.

34 Steven D. Levitt, "Understanding Why Crime Fell in the 1990s: Four Factors that Explain the Decline and Six that Do Not," *Journal of Economic Perspectives* 18, no. 1 (2004): 163–90, http://pricetheory.uchicago.edu/levitt/Papers/LevittUnderstandingWhyCrime2004.pdf.

35 William K. Rashbaum, "Retired Officers Raise Questions on Crime Data," *New York Times*, February 6, 2010, http://www.nytimes.com/2010/02/07/nyregion/07crime.html.

36 Rosabeth Moss Kanter and Stanley S. Litow, "Informed and Interconnected: A Manifesto for Smarter Cities," Working Paper 09-141, Harvard Business School, 2009, http://www.hbs.edu/research/pdf/09-141.pdf.

37 Tina Rosenberg, "Armed With Data, Fighting More Than Crime," *New York Times*, *Opinionator*, blog, last modified May 2, 2012, http://opinionator.blogs.nytimes.com/2012/05/02/armed-with-data-fighting-more-than-crime/.

38 Kanter and Litow, "Informed and Interconnected," 16.

39 Tolva, interview, November 10, 2011.

40 Tolva, interview, November 10, 2011.

41 Joe Flood, *The Fires* (Riverhead Books: New York, 2010), 207.

42 Katharine Q. Seeyle, "Menino to End Long Run as Boston Mayor," The Caucus blog, *New York Times*, March 28, 2013, http://thecaucus.blogs.nytimes.com/2013/03/28/menino-to-end-long-run-as-boston-mayor-reports-say/.

43 Thomas De Monchaux, "The Other Modernism," *N+1 Magazine*, July 12, 2012, http://nplusonemag.com/the-other-modernism.

44 Mike Barnicle, "Tom Menino, Urban Mechanic," *Boston Globe Magazine*, November 7, 1993, 29.

45 "Mayor Menino Invites Residents to 'Adopt-A-Hydrant' this Winter," City of Boston.gov, last modified January 19, 2012, http://www.cityofboston.gov/news/default.aspx?id=5444.

46 Nigel Jacob, telephone interview by author, August 13, 2012.

47 Nigel Jacob and Chris Osgood, telephone interview by author, March 25, 2011.

48 Jacob, interview, August 13, 2012.

49 Jacob and Osgood, interview, March 25, 2011.

50 Jacob, interview, August 13, 2012.

51 Jacob and Osgood, interview, March 25, 2011.

52 Jacob and Osgood, interview, March 25, 2011.

53 Jacob, interview, August 13, 2012.

54 Jacob and Osgood, interview, March 25, 2011.

55 Jacob and Osgood, interview, March 25, 2011.

56 All quotes in this section are from Daniel Sarasa and Juan Pradas, interview by author, November 30, 2011.

57 Anthony M. Townsend, "Digitally mediated urban space: new lessons for design," *Praxis: Journal of writing + building* 6 (2004), 100–105.

58 Steve Hamm, "Living Blogging from Smarter Cities Rio: Day 1," last modified November 9, 2011, http://asmarterplanet.com/blog/2011/11/live-blogging-from-smarter-cities-rio-day-1.html#more-12843.

59 Parag Khanna and David Skilling, "Big ideas from small places," *CNN Global Public Square*, blog, last modified November 1, 2011, http://globalpublicsquare.blogs.cnn.com/2011/11/01/big-ideas-from-small-places/.

60 Jacob and Osgood, interview, March 25, 2011.

61 Daniel Kaplan, "Open Public Data: Then What?—Part 1," *Open Knowledge Foundation*, blog, last modified January 11, 2008, http://blog.okfn.org/2011/01/28/open-public-data-then-what-part-1/.

Chapter 8. A Planet of Civic Laboratories

1 Peter Hirshberg, interview by author, November 15, 2011.

2 Quotes in this section from Peter Hirshberg, lecture, Technology Horizons Exchange, Institute for the Future, Sausalito, CA, October 26, 2011.

3 Jay Nath, lecture, Technology Horizons Exchange, Institute for the Future, Sausalito, CA, October 26, 2011.

4 Nath, Technology Horizons Exchange.

5 Peter Hirshberg, lecture, Technology Horizons Exchange, Institute for the Future, Sausalito, CA, October 26, 2011.

6 Nath, Technology Horizons Exchange.

7 "Urban Prototyping: Open Call," Gray Area Foundation for the Arts, http://sf.urbanprototyping.org/open-call/ accessed September 25, 2011.

8 Clay Shirky, "Situated Software," first published March 30, 2004 on the "Networks, Economics, and Culture" mailing list, http://www.shirky.com/writings/situated_software.html.

9 Shirky, "Situated Software."

10 "crowdSOS - Safety, Openness, and Security," n.d., accessed June 17, 2013, https://www.newschallenge.org/open/open-government/submission/crowdsos-safety-openness-and-security/.

11 Shirky, "Situated Software."

12 Kristen Purcell, Roger Entner, and Nichole Henderson, "The Rise of Apps Culture," Washington, DC: Pew Research Center, 2010, http://www.pewinternet.org/Reports/2010/The-Rise-of-Apps-Culture.aspx.

13 "TriMet App Center," n.d., http://trimet.org/apps/.

14 Patrick Geddes, *Civics as Applied Sociology* (Middlesex, UK: The Echo Library, 2008), 71.

15 Code for America, *2011 Annual Report*, San Francisco, 2011, http://codeforamerica.org/2011-annual-report/.

16 C. Shepard, "Code for America," *Urban Omnibus*, last modified August 11, 2010, http://urbanomnibus.net/2010/08/code-for-america/.

17 Jennifer Pahlka, telephone interview by author, January 18, 2012.

18 Nigel Jacob, telephone interview by author, August 13, 2012.

19 *Boston Globe*, "School Assignment Series," various dates, http://www.boston.com/news/education/specials/school_chance/articles/.

20 Jacob, interview, August 13, 2012.

21 Jacob, interview, August 13, 2012.

22 Tim O'Reilly, "Government as a Platform," 2010, http://ofps.oreilly.com/titles/9780596804350/.

23 Pahlka, interview, January 18, 2012.

24 Nigel Jacob and Chris Osgood, telephone interview by author, March 25, 2011.

25 Pahlka, interview, January 18, 2012.

26 Andrew Stevens and Jonas Schorr, "Reforming the world's city networks, Part 1: a time to cull," *Global Urbanist*, last modified April 11, 2012, http://globalurbanist.com/2012/04/11/city-networks.

27 Code for America, *2011 Annual Report.*

28 Sascha Haselmayer quotes from interview with author, February 8, 2011

29 City of Stockholm, "Accessibility," last modified January 10, 2012, http://international.stockholm.se/Press-and-media/Stockholm-stories/Accessibility/.

30 Aida Esteban, Sascha Haselmayer, and Jakob H. Rasmussen, *Connected Cities: Your 256 Billion Euro Dividend: How Innovation in Services and Mobility Contributes to the Sustainability of Our Cities* (London: Royal College of Art, 2010).

31 Sascha Haselmayer, interview with author, November 29, 2011.

32 William J. Clinton, lecture, World Business Forum 2011, New York, October 5, 2011.

33 Author's calculation based on published open-data sets and current US Census Bureau population estimates.

34 "City Protocol Framework", n.d., http://cityprotocol.org/framework.html.

35 Urban Systems Symposium: Defining Urban Systems, New York City, May 12, 2011.

36 Bertrand Russell, radio address, January 9, 1949, BBC Home Service, transcript at http://downloads.bbc.co.uk/rmhttp/radio4/transcripts/1948_reith3.pdf.

37 Eran Ben-Joseph, *The Code of the City: Standards and the Hidden Language of Place Making* (Cambridge, MA: MIT Press, 2005), 1.

Chapter 9. Buggy, Brittle, and Bugged

1 J. Casale, "The Origin of the Word 'Bug,'" *The OTB* (Antique Wireless Association), February 2004, reprinted at http://www.telegraph-history.org/bug/index.html.

2 Thomas P. Hughes, *American Genesis: A History of the American Genius for Invention* (New York: Penguin Books, 1989), 75.

3 William Maver Jr. and Minor M. Davis, *The Quadruplex* (New York: W. J. Johnston, 1890), 84.

4 http://www.history.navy.mil/photos/images/h96000/h96566k.jpg.

5 Kathleen Broome Williams, *Grace Hopper: Admiral of the Cyber Sea* (Annapolis, MD: Naval Institute Press, 2004), 54.

6 "Surge Caused Fire in Rail Car," *Washington Times*, last modified April 12, 2007, http://www.washingtontimes.com/news/2007/apr/12/20070412-104206-9871r/.

7 "About recent service interruptions, what we're doing to prevent similar problems in the future," Bay Area Rapid Transit District, last modified April 5, 2006, http://www.bart.gov/news/articles/2006/news20060405.aspx.

8 "The Economic Impact of Interrupted Service," *2010 U.S. Transportation Construction Industry Profile* (Washington, DC: American Road & Transportation Builders Association, 2010), http://www.artba.org/Economics/Econ-Breakouts/04_EconomicImpactInterruptedService.pdf.

9 Quentin Hardy, "Internet Experts Warn of Risks in Ultrafast Networks," *New York Times*, November 13, 2011, B3.

10 Ellen Ullman, "Op-Ed: Errant Code? It's Not Just a Bug," *New York Times*, last modified August 8, 2012, http://www.nytimes.com/2012/08/09/opinion/after-knight-capital-new-code-for-trades.html.

11 Charles Perrow, *Normal Accidents: Living with High-Risk Technologies* (Princeton, NJ: Princeton University Press, 1999), 4.

12 Robert L. Mitchell, "Y2K: The good, the bad and the ugly," *Computerworld*, last modified December 28, 2009, http://www.computerworld.com/s/article/9142555/Y2K_The_good_the_bad_and_the_crazy?taxonomyId=14.

13 David Green, "Computer Glitch Summons Too Many Jurors," *National Public Radio*, May 3, 2012, http://www.npr.org/2012/05/03/151919620/computer-glitch-summons-too-many-jurors.

14 Wade Roush, "Catastrophe and Control: How Technological Disasters Enhance Democracy," PhD diss., Program in Science, Technology and Society, Massachusetts Institute of Technology, 1994, http://hdl.handle.net/1721.1/28134.

15 Peter Galison, "War Against the Center," *Grey Room*, no. 4 (2001): 26.

16 Paul Baran, *On Distributed Communications* (RAND: Santa Monica, CA, 1964), document no. RM-3420-PR.

17 Barry M. Leiner et al., "Brief History of the Internet", n.d., http://www.internetsociety.org/internet/internet-51/history-internet/brief-history-internet, accessed August 29, 2012. It was the First ACM Symposium on Operating Systems Principles, http://dl.acm.org/citation.cfm?id=800001&picked=prox&CFID=171498151&CFTOKEN=24841121.

18 Bob Taylor, October 6, 2004, e-mail to Dave Farber reposted to INTERESTING-PEOPLE listserv, http://www.interesting-people.org/archives/interesting-people/200410/msg00047.html.

19 The 1977 geographical map of ARPANET, originally published in F. Heart, A. McKenzie, J. McQuillian, and D. Walden, *ARPANET Completion Report*, Bolt, Beranek and Newman, Burlington, MA, January 4, 1978, can be found at http://som.csudh.edu/fac/lpress/history/arpamaps/f15july1977.jpg.

20 "The Launch of NSFNET," n.d., http://www.nsf.gov/about/history/nsf0050/internet/launch.htm.

21 Marjorie Censer, "After Dramatic Growth, Ashburn Expects Even More

Data Centers," *Washington Post*, August 27, 2011, http://www.washington post.com/business/capitalbusiness/after-dramatic-growth-ashburn-expects -even-more-data-centers/2011/06/09/gIQAZduLjJ_story.html.

22 Steven Branigan and Bill Cheswick, "The effects of war on the Yugoslavian Network," 1999, http://cheswick.com/ches/map/yu/index.html.

23 William J. Mitchell and Anthony M. Townsend, "Cyborg Agonistes," in *The Resilient City: How Modern Cities Recover From Disaster*, edited by Lawrence J. Vale and Thomas J. Campanella (New York: Oxford University Press, 2005), 320–21.

24 New York State Public Service Commission, unpublished documents provided to the author.

25 Martin Fackler, "Quake Area Residents Turn to Old Means of Communication to Keep Informed," *New York Times*, March 28, 2011, A11.

26 National Research Council, Computer Science and Telecommunications Board, *The Internet Under Crisis Conditions: Learning From September 11* (Washington, DC: National Academies Press, 2003).

27 "Summary of the Amazon EC2 and Amazon RDS Service Disruption," last modified April 29, 2011, http://aws.amazon.com/message/65648/.

28 Chloe Albanesius, "Amazon Blames Power, Generator Failure for Outage," *PCMag.com*, July 3, 2012, http://www.pcmag.com/article2/0,2817,2406682 ,00.asp.

29 Christina DesMarais, "Amazon Cloud Hit by Real Clouds, Downing Netflix, Instagram, Other Sites," *Today @ PCWorld*, blog, June 30, 2012, http:// www.pcworld.com/article/258627/amazon_cloud_hit_by_real_clouds _knocking_out_popular_sites_like_netflix_instagram.html.

30 J. R. Raphael, "Gmail Outage Marks Sixth Downtime in Eight Months," *Today @ PCWorld*, blog, February 24, 2009, http://www.pcworld.com/article /160153/gmail_outage_marks_sixth_downtime_in_eight_months.html.

31 Author's calculation based on statistics reported in Massoud Amin, "U.S. Electrical Grid Gets Less Reliable," *IEEE Spectrum*, January 2011, http://spectrum .ieee.org/energy/policy/us-electrical-grid-gets-less-reliable.

32 Massoud Amin, "The Rising Tide of Power Outages and the Need for a Stronger and Smarter Grid," *Security Technology*, blog, Technological Leadership Institute, University of Minnesota, last modified October 8, 2010, http:// tli.umn.edu/blog/security-technology/the-rising-tide-of-power-outages-and -the-need-for-a-smart-grid/.

33 Maurice Gagnaire et al., "Downtime statistics of current cloud solutions," International Working Group on Cloud Computing Resiliency website, n.d.,

accessed February 14, 2013, http://iwgcr.org/wp-content/uploads/2012/06/IWGCR-Paris.Ranking-002-en.pdf.

34 Kathleen Hickey, "DARPA: Dump Passwords for Always-on Biometrics," *Government Computer News*, March 21, 2012, http://gcn.com/articles/2012/03/21/darpa-dump-passwords-continuous-biometrics.aspx.

35 Global Positioning System: Significant Challenges in Sustaining and Upgrading Widely Used Capabilities (US Government Accountability Office: Washington, DC), GAO-09-670T, May 7, 2009, http://www.gao.gov/products/GAO-09-670T.

36 *Global Navigation Space Systems: Reliance and Vulnerabilities* (London: Royal Academy of Engineering, 2011), 3.

37 "Scientists Warn of 'Dangerous Over-reliance on GPS,'" *The Raw Story*, March 8, 2011, http://www.rawstory.com/rs/2011/03/08/scientists-warn-of-dangerous-over-reliance-on-gps/.

38 "BufferBloat: What's Wrong with the Internet?" *ACMQueue*, blog, December 7, 2011, http://queue.acm.org/detail.cfm?id=2076798.

39 Jim Gettys and Kathleen Nichols, "Bufferbloat: Dark Buffers in the Internet," *ACMQueue*, blog, November 29, 2011, http://queue.acm.org/detail.cfm?id=2071893.

40 Ellen Nakashima and Joby Warrick, "Stuxnet was work of U.S. and Israeli experts, officials say," *Washington Post*, June 1, 2012, http://articles.washingtonpost.com/2012-06-01/world/35459494_1_nuclear-program-stuxnet-senior-iranian-officials.

41 Vivian Yeo, "Stuxnet infections spread to 115 countries," *ZDNet*, August 9, 2010, http://www.zdnet.co.uk/news/security-threats/2010/08/09/stuxnet-infections-spread-to-115-countries-40089766/.

42 Elinor Mills, "Ralph Langer on Stuxnet, copycat threats (Q&A)," CNet News, May 22, 2011, http://news.cnet.com/8301-27080_3-20061256-245.html.

43 Symantec Corporation, "W32.Stuxnet," *Security Responses*, blog, last modified September 17, 2010, http://www.symantec.com/security_response/writeup.jsp?docid=2010-071400-3123-99.

44 Dan Goodin, "FBI: No evidence of water system hack destroying pump," *The Register*, last updated November 23, 2011, http://www.theregister.co.uk/2011/11/23/water_utility_hack_update/.

45 Dan Goodin, "Rise of 'forever day' bugs in industrial systems threatens critical infrastructure," *Ars Technica*, April 9, 2012, http://arstechnica.com/business/news/2012/04/rise-of-ics-forever-day-vulnerabiliities-threaten-critical-infrastructure.ars.

46 Ellen Nakashima, "Cyber-intruder sparks massive federal response—and debate over dealing with threats," *Washington Post*, December 8, 2011, http://www.washingtonpost.com/national/national-security/cyber-intruder-sparks-response-debate/2011/12/06/gIQAxLuFgO_story.html.

47 Mark Ward, "Warning Over Medical Implant Attacks," BBC News, April 10, 2012, http://www.bbc.co.uk/news/technology-17623948; Daniel Halperin et al., "Pacemakers and Implantable Cardiac Defibrillators: Software Radio Attacks and Zero-Power Defenses," n.d., http://www.secure-medicine.org/icd-study/icd-study.pdf.

48 Colin Harrison, interview by author, May 9, 2011.

49 Chul-jae Lee and Gwang-li Moon, "Incheon Airport cyberattack traced to Pyongyang," *Korea JoongAng Daily*, June 5, 2012, http://koreajoongangdaily.joinsmsn.com/news/article/article.aspx?aid=2953940.

50 David E. Sanger, "Obama Order Sped Up Wave of Cyberattacks Against Iran," *New York Times*, June 1, 2012, A1.

51 Electronic Frontier Foundation, n.d., http://w2.eff.org/Privacy/TIA/wyden-sa59.php.

52 Alasdair Allan, "Got an iPhone or 3G iPad? Apple is recording your moves," *O'Reilly Radar*, April 20, 2011, http://radar.oreilly.com/2011/04/apple-location-tracking.html.

53 Trevor Eckhart, "CarrierIQ," *Android Security Test*, blog, n.d., http://androidsecuritytest.com/features/logs-and-services/loggers/carrieriq/.

54 Annalyn Censky, "Malls track shoppers' cell phones on Black Friday," *CNN Money*, blog, last modified November 22, 2011, http://money.cnn.com/2011/11/22/technology/malls_track_cell_phones_black_friday/index.htm.

55 Sebastian Anthony, "Think GPS is Cool? IPS Will Blow Your Mind," *ExtremeTech*, blog, last modified April 24, 2012, http://www.extremetech.com/extreme/126843-think-gps-is-cool-ips-will-blow-your-mind.

56 Timothy P. McKone, letter to Congressman Edward J. Markey, US House of Representatives, May 29, 2012, http://markey.house.gov/sites/markey.house.gov/files/documents/AT%26T%20Response%20to%20Rep.%20Markey.pdf.

57 Eric Lichtblau, "More Demands on Cell Carriers in Surveillance," *New York Times*, last modified July 8 2012, http://www.nytimes.com/2012/07/09/us/cell-carriers-see-uptick-in-requests-to-aid-surveillance.html?_r=1.

58 Loretta Chao and Don Clark, "Cisco Poised to Help China Keep an Eye on Its Citizens," *Wall Street Journal*, July 5, 2011, http://online.wsj.com/article/SB10001424052702304778304576377141077267316.html. Because of the way Chongqing's municipal boundaries are set, its population is often grossly

overstated. For discussion of urban area population estimate for Chongqing: Ruth Alexander, "The World's Biggest Cities, How Do You Measure Them," *BBC News Magazine*, last modified January 28, 2012, http://www.bbc.co.uk/news/magazine-16761784.

59 Clive Norris, Mike McCahill, and David Wood, "Editorial. The Growth of CCTV: A global perspective on the international diffusion of video surveillance in publicly accessible space," *Surveillance & Society*, http://www.surveillance-and-society.org/articles2(2)/editorial.pdf, 2(2/3): 110.

60 John Villasenor, *Recording Everything: Digital Storage as an Enabler of Authoritarian Governments* (Washington, DC: The Brookings Institution, 2011), http://www.brookings.edu/%7E/media/Files/rc/papers/2011/1214_digital_storage_villasenor/1214_digital_storage_villasenor.pdf, 1.

61 Chao and Clark, "Cisco Poised to Help China Keep an Eye on Its Citizens."

62 "Beijing to trial mobile tracking system: report," Agence France Presse, March 3, 2011.

63 David Goldman, "Carrier IQ: 'We're as surprised as you,'" *CNNMoney Tech*, blog, last modified December 2, 2011, http://money.cnn.com/2011/12/02/technology/carrier_iq/index.htm.

64 Farhad Manjoo, "Fear Your Smartphone," *Slate*, December 2, 2011, http://www.slate.com/articles/technology/technology/2011/12/carrier_iq_it_s_totally_rational_to_worry_that_our_phones_are_tracking_everything_we_do_.html.

65 Kate Notopoulos, "Somebody's watching: how a simple exploit lets strangers tap into private security cameras," *The Verge*, February 3, 2012, http://www.theverge.com/2012/2/3/2767453/trendnet-ip-camera-exploit-4chan.

66 Nicholas G. Garaufis, Memorandum & Order 10-MC-897 (NGG), August 22, 2011, http://ia600309.us.archive.org/33/items/gov.uscourts.nyed.312774/gov.uscourts.nyed.312774.6.0.pdf.

67 George Orwell, *1984* (Penguin: New York, 1990), 65.

68 Chao and Clark, "Cisco Poised to Help China Keep an Eye on Its Citizens."

69 Siobhan Gorman, "NSA's Domestic Spying Grows As Agency Sweeps Up Data," *Wall Street Journal*, March 10, 2008, http://online.wsj.com/article/SB120511973377523845.html.

70 John Villasenor, "Recording Everything: Digital Storage as an Enabler of Authoritarian Governments" (Washington, DC: Brookings Institution, December 14, 2011), 1.

71 Herman Kahn, *Thinking About the Unthinkable* (New York, Horizon Press, 1962).

72 "How U.S. Cities Can Prepare for Atomic War," *Life*, December 18, 1950, 85.

73 Light, *From Warfare to Welfare*, 164.

74 Galison, "War Against the Center," 14–26.

75 *World Energy Outlook 2011* (Paris: International Energy Agency, 2011).

76 *Realizing the Potential of Energy Efficiency: Targets, Policies, and Measures for G8 Countries* (Washington, DC: United Nations Foundation, 2007), http://www.globalproblems-globalsolutions-files.org/unf_website/PDF/realizing_potential_energy_efficiency.pdf.

77 Buno Berthon, "Smart Cities: Can They Work?," *The Guardian Sustainable Business Energy Efficiency Hub*, blog, June 1, 2001, http://www.guardian.co.uk/sustainable-business/amsterdam-smart-cities-work.

78 Blake Alcott, "Jevons' Paradox," *Ecological Economics* 45, no. 1 (2005): 9-21.

79 Robert Cervero, *The Transit Metropolis* (Washington, DC: Island Press, 1998), 169.

80 Michele Dix, "The Central London Congestion Charging Scheme—From Conception to Implementation," 2002, http://www.imprint-eu.org/public/Papers/imprint_Dix.pdf, 2.

81 Robert J. Gordon, "Does the 'New Economy' Measure up to the Great Inventions of the Past?" (Cambridge, MA: National Bureau of Economic Research, 2000), http://www.nber.org/papers/w7833.

Chapter 10. A New Civics for a Smart Century

1 Oscar Wilde, *The Soul of Man under Socialism* (Portland, ME: Thomas B. Mosher, 1905), 39. Reprinted from *The Fortnightly Review*, Feburary 1, 1891, accessed through Internet Archive, http://archive.org/details/soulmanundersoc00wildgoog.

2 Helen Meller, *Patrick Geddes: Social Evolutionist and City Planner* (New York: Routledge, 1990), 143.

3 From "voices to voices, lip to lip." Copyright 1926, 1954, © 1991 by the Trustees for the E. E. Cummings Trust. Copyright © 1985 by George James Firmage, from *Complete Poems: 1904–1962* by E. E. Cummings, edited by George J. Firmage. Used by permission of Liveright Publishing Corporation.

4 Brandon Fuller and Paul Romer, "Success and the City: How Charter Cities Could Transform the Developing World," (Ottawa, Ontario: The MacDonald-Laurier Institute, April 2012), 3.

5 William J. Mitchell, *E-Topia: Urban Life, Jim, But Not As We Know It* (Cambridge, MA: MIT Press, 1999), 12.

6 Jon Leibowitz, "Municipal Broadband: Should Cities Have a Voice?" National Association of Telecommunications Officers and Advisors (NATOA) 25th Annual Conference, Washington, DC, September 22, 2005, http://www.ftc.gov/speeches/leibowitz/050922municipalbroadband.pdf.

7 Christopher Mitchell, *Broadband at the Speed of Light: How Three Communities Build Next-Generation Networks* (Washington, DC: Institute for Local Self-Reliance, April 2012, http://www.ilsr.org/wp-content/uploads/2012/04/muni-bb-speed-light.pdf.

8 Dave Flessner, "Chattanooga area's economic outlook brightens," *Chattanooga Times Free Press*, last modified December 29th, 2011, http://www.timesfreepress.com/news/2011/dec/29/economic-outlook-brightens.

9 Fiber-to-the-Home Council of North America, "Municipal Fiber to the Home Deployments: Next Generation Broadband as a Municipal Utility," October 2009, http://www.baller.com/pdfs/MuniFiberNetsOct09.pdf.

10 Claudia Sarrocco and Dimitri Ypsilanti, "Convergence and Next Generation Networks: Ministerial Background Report," Organisation for Economic Co-operation and Development, June 17, 2008, http://www.oecd.org/internet/interneteconomy/40761101.pdf.

11 Christopher Mitchell, "Oregon Town To Build Open Access Fiber Network Complement to Wireless Network," Community Broadband Networks, last modified July 25, 2011, http://www.muninetworks.org/content/oregon-town-build-open-access-fiber-network-complement-wireless-network.

12 Thierry Martens, remarks, Ideas Economy: Intelligent Infrastructure, *The Economist*, New York City, February 16, 2011.

13 Andrew Comer and Kerwin Datu, "Can you have a private city? The political implications of 'smart city' technology," *Global Urbanist*, last modified February 11, 2011, http://globalurbanist.com/2011/02/17/can-you-have-a-private-city-the-political-implications-of-smart-city-technology.

14 Jennifer Pahlka, panel discussion, *Ten Year Forecast Retreat*, Institute for the Future, Sausalito, CA, April 15, 2012.

15 Carlo Ratti, lecture, Forum on Future Cities, MIT SENSEable City Lab and the Rockefeller Foundation, Cambridge, MA, April 12, 2011, http://techtv.mit.edu/collections/senseable/videos/12257-smart-smarter-smartest-cities.

16 IBM Corp., "Citi Partners with Streetline and IBM to Provide $25 Million Financing for Cities to Adopt Smart Parking Technology," last modified April 9, 2012, http://www-03.ibm.com/press/us/en/pressrelease/37424.wss.

17 Eve Batey, "Muni App Makers, Rejoice: MTA, Apple Disputes Private Company's Claims To Own Arrival Data," *SF Appeal Online Newspaper*, last modified August 19, 2009, http://sfappeal.com/news/2009/08/mike-smith-of-nextbus-said.php.

18 Joe Mullin, "A New Target for Tech Patent Trolls: Cash-Strapped American Cities," *Ars Technica*, last modified March 15, 2012, http://arstechnica.com/tech-policy/2012/03/a-new-low-for-patent-trolls-targeting-cash-strapped-cities/.

19 John Tolva, interview by author, November 10, 2011.

20 Steve W. Usselman, "Unbundling IBM: Antitrust and the Incentives to Innovation in American Computing," in Clarke, Lamoreaux, and Usselman, eds., *The Challenge of Remaining Innovative* (Palo Alto, CA: Stanford University Press, 2009), 251.

21 Dom Ricci, remarks, X-Cities 3: Heavy Weather—Design and Governance in Rio de Janeiro and Beyond, Columbia University Studio-X, New York, April 10, 2012.

22 Noelle Knell, "Detroit Pulls Plug on 311 Call Center," *Government Technology*, last modified July 11, 2012, http://www.govtech.com/e-government/Detroit-Pulls-Plug-on-311-Call-Center.html.

23 Michael Batty, "A Chronicle of Scientific Planning: The Anglo-American Modeling Experience," *Journal of the American Planning Association* 60, no. 1 (1994): 7.

24 Michael Batty, telephone interview by author, August 19, 2010.

25 Douglass B. Lee Jr., "Requiem for Large-Scale Models," *Journal of the American Institute of Planners* 39, no. 3 (1973): 173.

26 Michael Batty, lecture, "Forum on Future Cities," MIT SENSEable City Lab and the Rockefeller Foundation, Cambridge, MA, April 13, 2011, http://techtv.mit.edu/collections/senseable/videos/12305-changing-research.

27 David Weinberger, "The Machine That Would Predict the Future," *Scientific American*, November 15, 2011, http://www.scientificamerican.com/article.cfm?id=the-machine-that-would-predict.

28 Lee, "Requiem," 175.

29 Justin Cook, telephone interview by author, September 11, 2012.

30 David Gelernter, *Mirror Worlds: or the Day Software Puts the Universe in a Shoebox . . . How It Will Happen and What It Will Mean* (New York: Oxford University Press, 1993), 19.

31 Colin Harrison, interview by author, May 9, 2011.

32 Jay Nath, "Hacking SF: Innovation in Public Spaces," *Jay Nath*, blog, last modified April 12, 2012, http://www.jaynath.com/2012/04/hacking-sf-innovation-in-public-spaces/.

33 Phil Bernstein, remarks, Bill Mitchell Symposium, MIT Media Lab, Cambridge, MA, Nov 11, 2011.

34 "The Transect," Center for Applied Transect Studies, accessed September 5, 2012, http://www.transect.org/transect.html.

35 Red Burns, "Technology and the Human Spirit," lecture at "The Future of Interactive Communication," Lund, Sweden, June 1998.

36 "Transdisciplinarity," *Science and Technology Outlook: 2005–2055* (Palo Alto, CA: Institute for the Future, 2006), 31, http://www.iftf.org/system/files/deliverables/TH_SR-967_S%2526T_Perspectives.pdf.

37 Adam Greenfield, "Beyond the 'smart city,'" *Urban Scale*, blog, last modified February 17, 2011, http://urbanscale.org/news/2011/02/17/beyond-the-smart-city/.

38 Evgeny Morozov, "Technological Utopianism," *Boston Review*, November/December 2010, http://www.bostonreview.net/BR35.6/morozov.php.

39 Michael M. Grynbaum, "Mayor Warns of Pitfalls of Social Media," *New York Times*, March 21, 2012, http://www.nytimes.com/2012/03/22/nyregion/bloomberg-says-social-media-can-hurt-governing.html.

40 Italo Calvino, *Invisible Cities* (New York: Harcourt, 1974), 32.

41 Michael Joroff, e-mail correspondence with author, January 28, 2012.

42 Janette Sadik-Khan, lecture, "BitCity 2011: Transportation, Data and Technology in Cities," Columbia University, New York City, November 4, 2011.

43 Guru Banavar, lecture, "X-Cities 3: Heavy Weather—Design and Governance in Rio de Janeiro and Beyond," Columbia University Studio-X, New York, April 10, 2012, http://www.youtube.com/watch?v=xNsSNoL_EQM.

44 Frank Hebbert, interview by author, April 12, 2012.

45 Joroff, January 28, 2012.

46 Synopsis, *Ekumenopolis: City Without Limits*, 2012, film directed by İmre Azem, produced by Gaye Günay, http://www.ekumenopolis.net/#/en_US/synopsys, accessed September 19, 2012.

47 "Insights in Motion: Improving Public Transit," Official IBM Social Media Channel, last modified June 12, 2012, http://www.youtube.com/watch?v=KEpVJscv7qE .

48 Alexis de Tocqueville, *Democracy in America*, vol. 2, ch. 5, electronic edition by the American Studies Programs at the University of Virginia, (1997), http://xroads.virginia.edu/~Hyper/DETOC/ch2_05.htm.

49 Regional Plan Association, "Crowds: In the City There Are Always Crowds" (New York: Regional Plan Association, 1932).

50 Steven Johnson, "What 100 Million 311 Calls Reveal About New York,"

Wired, November 2010, http://www.wired.com/magazine/2010/11/ff_311_new_york/.

51 Robert Goodspeed, "The Democratization of Big Data," *Planetizen*, blog, last modified February 27, 2012, http://www.planetizen.com/node/54832.

52 Jonah Lehrer, "A Physicist Solves the City," *New York Times Magazine*, December 17, 2010, http://www.nytimes.com/2010/12/19/magazine/19Urban_West-t.html.

53 Cosma Rohilla Shalizi, "Scaling and Hierarchy in Urban Economies," April 8, 2011, http://arxiv.org/abs/1102.4101.

54 Michael Batty and Elsa Arcaute, Skype interview by author, October 19, 2012.

55 Batty and Arcaute, interview, October 19, 2012. The study was subsequently published on the arXiv e-print archive. Elsa Arcaute et al., "City boundaries and the universality of scaling laws," January 8, 2013, http://arxiv.org/abs/1301.1674.

56 Cosma Rohilla Shalizi, "Scaling and Hierarchy in Urban Economies," *ARXIV*, e-print arXiv:1102.4101, February 2011, http://arxiv.org/abs/1102.4101.

57 Steve Lohr, "SimCity, for Real: Measuring an Untidy Metropolis," *New York Times*, February 23, 2013, BU3.

58 Geoffrey West, lecture, Urban Systems Symposium, New York University, New York City, May 12, 2012.

59 "Thinking Cities: ICT is Changing the Game," Telefonaktiebolaget LM Ericsson, last modified February 24, 2012, http://www.ericsson.com/news/120221_thinking_cities_ict_is_changing_the_game_244159020_c.

60 Hirshberg, interview, October 26, 2011.

61 Michael Batty, interview, August 19, 2010.

62 William Bruce Cameron, *Informal Sociology: A Casual Introduction to Sociological Thinking* (New York: Random House, 1967) 13.

63 Upton Sinclair, *The Jungle* (New York: The Jungle Pub. Co., 1906), 67.

64 Elan Miller, "Redesigning Lost & Found," *Still Hungry, Still Foolish*, blog, last modified December 14, 2011, http://elanmiller.com/post/14214715871/redesigning-lost-found.

Index